The Mirror World

Harnessing the Power of Digital Twins

By
Michael Lawson

The Mirror World

Harnessing the Power of Digital Twins

Table of Contents

Introduction

The concept of digital twins is reshaping the way we understand and interact with the world around us. It represents more than just a technological advancement; it marks a new chapter in the union of the physical and the digital realms. As we stand at the precipice of this exciting frontier, technology enthusiasts, industry professionals, and decision-makers are coming together, excited yet inquisitive about the possibilities digital twins present.

At its core, a digital twin is a virtual representation of a physical object or system. However, the magic lies in its ability to continuously learn and update based on real-time data inputs. This dynamic mirroring is what sets digital twins apart from static models, allowing them to provide insights that were previously unimaginable. The potential applications are endless, touching every sector from manufacturing to healthcare, smart cities to agriculture, and beyond.

The journey toward understanding digital twins begins with a simple acknowledgment: these innovations aren't about replacing our physical experiences but enhancing them. They serve as a bridge, facilitating a seamless flow of information between the material and digital worlds. This fluid exchange empowers us to predict outcomes, optimize processes, and ultimately drive innovation across industries. Imagine understanding a manufacturing flaw before it halts an assembly line or tailoring medical treatments to the personal biology of each patient. These examples are no longer science fiction; they're becoming reality.

Yet, adopting digital twin technology isn't just about integrating new tools into existing frameworks. It requires a fundamental shift in mindset. Companies and individuals must embrace a culture of continuous learning and adaptation. The digital twin paradigm challenges traditional approaches, urging us to take bold steps in building interconnected ecosystems where data-driven decisions pave the way for transformative outcomes. This is where inspiration meets action, as we harness the power of digital twins to propel us into an era of unprecedented innovation.

Central to the appeal of digital twins is their ability to increase efficiency and reduce waste. Organizations can plan, test, and refine their processes in a virtual environment before implementing changes in the real world. This proactive approach minimizes risks and maximizes rewards, making digital twins a cornerstone of any strategy aimed at sustainable growth and long-term success. By visualizing complex systems in a simplified manner, digital twins allow us to see the unseen, anticipate issues before they arise, and solve problems with precision and agility.

But the impact of digital twins extends beyond individual organizations. They hold the potential to revolutionize entire industries, driving systemic changes that align with the broader goals of society. Whether enhancing urban planning in smart cities, optimizing energy systems to reduce environmental impact, or revolutionizing medical research with patient-centric innovations, digital twins offer the tools to build a better future for all. This technology encourages collaboration across disciplines, fostering an environment of shared knowledge and collective progress.

In navigating this path forward, the role of digital twins as catalysts for innovation cannot be overstated. They inspire us to push boundaries, explore new frontiers, and dream bigger than before. As we delve deeper into the nuances of this technology, we must continually ask ourselves how we can harness its potential to address the challenges facing our world today. This inquiry is not just academic; it is profoundly practical and deeply necessary.

While the journey to full-scale adoption may present challenges, its rewards are promising. For those prepared to embrace this digital evolution, a world of opportunities lies ahead. Digital twins are not merely a trend or a fleeting interest. They represent a fundamental shift in how we approach complex problems, a shift backed by data-driven insights and human ingenuity. Armed with the tools to make calculated decisions, organizations can maintain a competitive edge in an ever-changing landscape.

The time to act is now. By seeking to understand, adopt, and innovate with digital twin technology, we pave the way for a future where creativity meets precision, where our collective potential is harnessed for the greater good. As technology enthusiasts, professionals, and leaders, the onus is on us to ensure that this transition to a digital twin-enabled world is realized smoothly and sustainably. It's not just about being prepared for the future; it's about actively creating it.

Chapter 1: Understanding Digital Twins

Digital twins are quietly reshaping how industries approach innovation, embedding themselves as the keystone in the bridge between the physical and digital worlds. Think of them as the dynamic duet of real-world systems and their virtual counterparts, constantly exchanging data to refine and optimize performance. The term itself may sound like a futuristic concept, but its roots are firmly anchored in practical necessity, initially sprouting in the aerospace sector before branching out across various fields. In essence, digital twins simulate real-time scenarios enabling businesses to predict outcomes, troubleshoot issues, and enhance understanding. This symbiotic relationship between the digital and the tangible isn't just about mirroring physical objects; it's about fostering a learning loop where each side informs and refines the other. As we dive deeper into the realm of digital twins, it becomes evident that they hold the potential to push the boundaries of innovation, making our systems smarter, more efficient, and incredibly adaptive in an ever-evolving landscape.

Exploring the Basics

To truly grasp what digital twins are, we need to start with the fundamentals: a digital twin is essentially a virtual representation of a physical object, system, or process. But it's much more than a mere digital replica. Think of it as a bridge connecting the digital and physical worlds, allowing us to simulate, predict, and optimize real-world processes in a way that's both revolutionary and immensely practical. This concept leverages a myriad of technologies, including the Internet of Things (IoT), artificial intelligence (AI), and data analytics, to give us a robust framework for understanding reality like never before.

At its core, the digital twin paradigm is about creating a symbiotic relationship between physical entities and their virtual counterparts. This harmony allows us to emulate a wide range of scenarios right in the digital space, venturing into possibilities that simply wouldn't exist otherwise. The power lies not only in monitoring and analyzing but also in predicting future behaviors and outcomes. This capability opens up a whole new level of insights and foresight for various industries, from manufacturing to healthcare, urban planning, and beyond.

The concept of a digital twin might seem cutting-edge, but it builds on the idea of paired systems that have been in development for decades. Previously, engineers relied on Computer-Aided Design (CAD) and similar technologies to create static models for analysis and design. However, digital twins push these boundaries further by integrating real-time data, allowing continuous updates and iterative learning based on real-world inputs. In doing so, digital twins become dynamic and living models that adapt as the physical object evolves.

Understanding the basic mechanics of digital twins starts with recognizing the loop of continuous interaction between the physical and digital. This interaction involves the collection of data from sensors and other inputs, which is then fed into the digital model. The model processes this data, providing insights, forecasts, and recommendations back to the physical counterpart. This cyclical process is instrumental in improving efficiency, reducing costs, and driving innovation.

One might wonder, how does this tie into everyday applications? Consider a smart factory: here, digital twins monitor machinery performance, predict failures before they happen, and optimize production lines for maximum efficiency. These virtual models transform data into actionable insights, enhancing decision-making with precision. The necessity for this kind of real-time problem-solving grows exponentially in today's fast-paced industrial landscape.

Moreover, digital twins aren't confined to large-scale industrial applications; they're just as valuable in personalizing consumer experiences. For instance, in smart homes, digital twins can optimize energy use by modeling household behaviors and adjusting heating or lighting autonomously. This kind of tailor-made interaction elevates the promise of digital twin technology from just a futuristic concept to a practical tool that enhances our daily lives.

What sets digital twins apart from other technological advancements is their ability to learn over time. Through machine learning algorithms, these virtual entities can evolve with the systems they replicate. They don't just mirror the present; they create a feedback loop that informs future decisions, making them proactive rather than reactive. The application of machine learning within digital twins facilitates the understanding of complex patterns, trends, and anomalies that would be nearly impossible to determine manually.

To fully unravel the basics of digital twins, it's essential to appreciate the role of data. Data is the lifeblood of digital twins, enabling real-time analysis and allowing them to inform and guide physical counterparts. But it's not just any data—it's high-quality, relevant data sourced from numerous points: sensors, business systems, and environmental inputs. The richness and accuracy of this data directly influence the effectiveness of a digital twin, underscoring the significance of robust data management practices in this realm.

With the groundwork now laid, one might ask how digital twins fit into the broader spectrum of digital transformations occurring worldwide. The truth is, they're at the heart of it, representing a convergence of the latest technologies that are reshaping industries. As we continue to navigate the digital age, it's vital to embrace digital twins as a cornerstone for innovation; they're not merely a technological tool, but a strategic driver of change that holds the potential to redefine business models and operational paradigms.

In many ways, unlocking the potential of digital twins requires a mindset shift more than a technological one. It's about rethinking processes and questioning the status quo, challenging traditional inefficiencies, and exploring paths previously deemed impossible. As technology enthusiasts and industry professionals, diving deep into the intricacies of digital twins isn't just about understanding their functionality—it's about recognizing their transformative potential to inspire new frontiers in innovation.

In summary, digital twins are more than a trend; they represent a digital evolution in which technology and real-world systems coalesce to provide unmatched insight, efficiency, and creativity. By exploring the basics, we've begun to unlock an understanding of their fundamental workings and realized the impact they can have across diverse sectors. As we continue to explore and innovate with digital twins, we pave the way for unprecedented advancements, turning science fiction dreams into tangible, real-world solutions.

Historical Context and Evolution

The idea of creating a digital representation of a physical system dates back to concepts and theories from the early 20th century. Visionaries and engineers have long been intrigued by the notion of using technology to mimic and understand complex systems. Early efforts were rudimentary, but they laid the groundwork for what we now call digital twins. These nascent technologies focused on simple simulations and models, primarily for engineering and scientific purposes. Though limited in scope, they inspired future generations to explore in more detail the intricate relationship between the physical and digital realms.

The real momentum for digital twins began with the advent of computer-aided design (CAD) systems in the 1960s and 1970s. CAD systems revolutionized how engineers and designers approached product development. Whereas before, physical prototypes or limited mechanical drawings were used, digital models allowed for iterative and detailed design without expensive physical trials. This shift represented a fundamental change and planted the first seeds of digital twin technology, where a digital counterpart could be tested, manipulated, and improved upon before any physical iteration existed.

By the late 20th century, the rapid advancement in computing power and software capabilities enabled more complex simulations and virtual modeling environments. Industries like aerospace and automotive increasingly moved from simple CAD models to advanced virtual representations that incorporated multiple variables and conditions. These environments got closer to real-world conditions and were a precursor to full-fledged digital twins. The need for precise simulations that could predict behavior under different circumstances fostered innovation, driving digital twin development further.

In the early 2000s, digital twins took a major leap with the rise of smart sensors and IoT technologies. Suddenly, it was possible to have real-time data streaming from physical objects into their digital counterparts. This real-world data could be used to update digital models in real-time, allowing for unprecedented accuracy and insight. GE was among the pioneers in recognizing this potential, using digital twins to optimize the performance and maintenance schedules of industrial turbines. This marked a shift from static models to dynamic entities that changed and evolved alongside their physical counterparts.

Digital twins continued to mature with the integration of artificial intelligence and machine learning in the 2010s. These technologies provided the tools to analyze massive datasets generated by twin systems. Predictive analytics became possible, and insights could be gleaned automatically, transforming decision-making processes across industries. AI allowed digital twins to move beyond mere reflections of physical systems to becoming predictive tools that could anticipate failures and optimize operations. This evolution aligned closely with increasing digitization across industries, creating a fertile ground for new applications and developments.

The expansion of digital twin technology into new sectors parallels the ongoing digital transformation businesses everywhere are experiencing. From smart cities leveraging digital twins for urban planning to healthcare systems employing them for patient-specific treatments, the technology's applications have grown exponentially. Each implementation brings unique challenges and innovations, further refining the core concept and expanding its potential uses.

Industry 4.0 technologies have further shaped and refined digital twin applications, pushing their capabilities beyond initial expectations. The theory behind Industry 4.0 speaks to interconnectedness, autonomous systems, and data-driven decision making. Digital twins fit perfectly within this paradigm, offering a digital reflection of physical reality which, when combined with other technologies like blockchain and augmented reality, provides comprehensive oversight and control over complex systems.

Continuous improvements in data processing, storage capabilities, and network speeds have further spurred digital twin development. Cloud computing now plays a crucial role, offering scalable resources to handle the tremendous data volumes that digital twins necessitate.

The synergy between cloud infrastructures and digital twin solutions allows for easier management, better reliability, and enhanced accessibility. These improvements are crucial in fields where accuracy and real-time responsiveness are critical.

Today, as the technology is entering its golden era, digital twins are not merely theoretical constructs but operational necessities. Industries globally are innovating around this technology, pushing boundaries, and disrupting traditional practices with digital twins leading the charge. In doing so, they have unlocked a new dimension of interactivity and analytical depth that was previously unattainable. The technology has emerged as an enabler for innovation, offering pathways to holistic and informed decision-making by bridging the gap between digital and physical worlds.

As we gaze into the future, the evolution of digital twins promises even greater transformations. Emerging technologies such as quantum computing and 5G networks will likely propel digital twins to new heights, enabling richer models and faster computations. The relationship between digital twins and emerging fields like autonomous systems and edge computing offers fertile grounds for groundbreaking innovations that could redefine how industries operate and collaborate.

As digital twins continue to evolve, their role not just as mere reflections or analytical tools, but as integral components of the physical systems they represent, will become more pronounced. The journey from concept to indispensable technology underscores a deep commitment to innovation and exemplifies the transformative power of human ingenuity in leveraging emerging technologies for real-world impact. The path of digital twins is a testament to the synergy between technological progress and the ever-expanding aspirations of industries worldwide.

Chapter 2:
The Architecture of Digital Twins

The architecture of digital twins is the beating heart that powers their extraordinary capabilities, weaving together a tapestry of technologies to create a comprehensive virtual replica of physical systems. At its core, this architecture intricately combines diverse components and technologies, creating a robust framework that drives seamless integration and functionality. The fusion of data analytics, modeling, and simulation allows these dynamic models to not only mimic but also predict the behavior of their real-world counterparts. What truly sets digital twins apart is their symbiotic relationship with cutting-edge technologies like the Internet of Things (IoT) and artificial intelligence. This synergy enables them to adapt, learn, and evolve over time, providing an unparalleled level of insight and foresight. As industries embrace digital twins, understanding their architecture becomes vital to unlocking exponential innovation potential, catapulting industries into a future where real-time decision-making merges with strategic foresight, and the line between the digital and physical worlds continues to blur.

Key Components and Technologies

The architecture of digital twins is a fascinating construct, a multi-layered platform which brings together technology, data, and analytics to create real-time virtual representations of physical systems. At its core, the architecture intertwines a web of key components and pioneering technologies. Each piece plays a critical role, like a note in a symphonic orchestra, creating a harmonious and dynamic structure capable of transforming insights into impactful actions.

Central to the functionality of digital twins are their ability to handle copious amounts of data. Sensors are the frontline components that facilitate this data acquisition. These sophisticated devices capture a continuous stream of information from the physical world, delivering it seamlessly to the digital counterpart. From temperature and pressure to motion and speed, sensors are the bridge enabling the fluid exchange of information between the physical and digital realms.

Yet, sensors alone can't sustain the digital twin's operations. The data they gather would be void without the right technologies for storage and processing. This's where data management systems come into play. They provide robust frameworks for collecting, storing, retrieving, and managing data. Structured and unstructured data flood these systems, and at their helm are databases designed to not only handle the data effectively but also to support analytics and visualizations.

Accompanying data management systems is connectivity technology that ensures data is transmitted accurately and in real-time. The advent of high-speed networks like 5G has significantly enhanced this capability. Fast and reliable connectivity isn't just a benefit but a necessity for digital twins to function optimally, as it supports the instantaneous flow and updating of data, allowing for real-time tracking and analysis.

The computational backbone of digital twins would be incomplete without analytics engines. These engines are crucial in transforming raw data into actionable insights. Through advanced algorithms and machine learning, analytics technologies delve deep into data pools, distinguishing patterns and making predictions. This analytical transformation enables organizations to optimize operations, forecast trends, and create scenarios for better decision-making.

In parallel, digital twin technology leverages visualization tools. By crafting intuitive and interactive 3D models, these tools translate complex data into comprehensible insights. This visual representation not only aids in monitoring and analyzing the system but also provides an intuitive medium for stakeholders to interact with the data, thereby making it easier to identify issues and opportunities at a glance.

Security mechanisms are another indispensable component. Given the sensitive nature of the data handled by digital twins, robust cybersecurity protocols are critical. These technologies implement protective measures to safeguard the integrity and privacy of data, employing encryption, access controls, and continuous monitoring to defend against potential threats.

Then there's the integration of simulation technologies, the essence of what makes a digital twin a true "twin." By recreating complex physical processes within a virtual environment, simulations allow for testing without risks, enabling optimizations and hypotheses testing before they're applied to the real world. This capability to preemptively troubleshoot and refine is a game-changer for industries ranging from aerospace to healthcare.

Underpinning these components is an architecture that's grounded in modular design. This modularity provides the flexibility to customize and scale digital twins according to specific needs or changes in scope. It supports the integration of new technologies and data sources as they evolve, ensuring that the digital twin remains relevant and capable of addressing contemporary challenges.

Interestingly, cloud computing has become a fundamental technology supporting the digital twin paradigm. Offering scalability, accessibility, and cost-efficiency, cloud platforms provide the infrastructure necessary for storing and processing vast amounts of data generated by digital twins. They facilitate the rapid deployment and management of digital twins across distributed environments, linking disparate data sources into a cohesive framework.

As we continue to explore the limitless potential of digital twins, hybrid integration platforms have surfaced as vital facilitators of seamless connectivity across multiple systems and technologies. By orchestrating data flow and functional integration, these platforms ensure data consistency and interoperability, driving efficiency and accelerating innovation.

With each component and technology building on the next, the architecture of digital twins forms a robust ecosystem. It's this intricate synergy that empowers organizations to gain deeper insights, improve performance, and catalyze innovation. As technologies advance, the architecture of digital twins will adapt and evolve, remaining at the forefront of digital transformation.

Integration with IoT and AI

The architecture of digital twins finds its most dynamic potential in the seamless integration with the Internet of Things (IoT) and Artificial Intelligence (AI). By marrying these technologies, digital twins transcend their static data representations and transform into intricate systems that mimic real-world entities and processes. This synergy creates a robust framework where real-time data and intelligent insights converge, fostering innovation and opening new avenues for both business and societal advancements.

At the heart of this integration is IoT, which acts as the sensory layer of digital twins. IoT devices continuously gather data from the physical world, forming a vital data stream into the digital twin models. From industrial machines on a factory floor to sensors in a smart city, IoT captures real-time data such as temperature, pressure, speed, and more. This influx of information feeds digital twins, ensuring they reflect the current state of their real-world counterparts accurately and reliably. In essence, IoT forms the nervous system of digital twins, enabling them to perceive and react to the physical world with unerring precision.

However, sensing the physical world is just one side of the coin. The real power lies in how this data is processed and analyzed. Here, AI steps in, offering computational capabilities that transform raw data into actionable insights. By leveraging machine learning algorithms, AI enables digital twins to recognize patterns, predict future states, and simulate scenarios. This intelligent processing allows digital twins to offer more than just a mirror image of their physical counterparts; they become insightful advisors, predicting failures, recommending optimizations, and even adapting autonomously to changing conditions.

The combination of IoT and AI within digital twins fosters a level of interactivity and adaptability that traditional models can't achieve. Consider the manufacturing sector, where a digital twin of a production line can monitor operations in real-time. IoT sensors detect an unexpected vibration in machinery, streaming this data to the digital twin. Through AI analysis, the twin might identify a potential mechanical failure and suggest preemptive maintenance to avoid costly downtime. This proactive approach enhances efficiency, reduces costs, and prevents disruptions, showcasing the tangible value of IoT and AI within digital twin ecosystems.

Smart cities provide another compelling case study for this integration. Digital twins of city infrastructures, enhanced by IoT sensors and AI, enable real-time monitoring and optimization of urban resources. Traffic patterns can be analyzed in real-time, offering insights to alleviate congestion and minimize emissions. Resource allocation, such as electricity and water, can be dynamically managed based on current demand, while predictive models can guide city planning and development. In this way, IoT and AI do more than enhance urban living; they revolutionize it, making cities more sustainable and responsive to the needs of their inhabitants.

In healthcare, digital twins supported by IoT and AI are paving the way for transformative patient-centric care. Wearable devices track patient health metrics, feeding this data into patient-specific twins. AI processes these continuous streams, allowing for early detection of anomalies and personalized treatment recommendations. For chronic conditions, this approach offers a significant improvement in patient outcomes by enabling proactive interventions rather than reactive care. Furthermore, by simulating patient responses to different treatments, healthcare providers can tailor interventions to individual needs, enhancing both efficacy and patient experience.

While the benefits of weaving IoT and AI into the fabric of digital twins are manifold, it also introduces challenges that require meticulous attention. The sheer volume of data generated by IoT sensors mandates robust data management strategies and infrastructure to handle this inflow efficiently. Ensuring data security and privacy is paramount, as the sensitivity of the information handled—whether personal health data, industrial secrets, or city infrastructure details—demands stringent protection measures.

Moreover, integrating AI into digital twins necessitates continuous model retraining and optimization to maintain accuracy and relevance. The algorithms must be imbued with the ability to evolve alongside technological advancements and changing environmental conditions. This dynamic modification underpins the longevity and utility of digital twins, ensuring they remain a step ahead in providing value-driven insights and predictions.

The fusion of IoT and AI into digital twins is more than a technological advancement; it's a paradigm shift, reshaping how industries operate and innovate. By building systems that not only reflect the current state but foresee future possibilities, businesses and cities can achieve levels of productivity, efficiency, and sustainability previously thought unreachable. The encouraging narrative here is that the possibilities are unlimited, as long as stakeholders are willing to embrace these technologies and evolve alongside their potential.

This harmonious integration doesn't just enhance the operational effectiveness of individual sectors. It also sets the stage for cross-disciplinary innovation that can address complex, multifaceted challenges facing modern society. Energy efficiency, climate change mitigation, public health improvement—these global concerns can be tackled more effectively when digital twins, powered by IoT and AI, are employed to provide holistic insights and coordinated action plans across multiple domains.

In summation, the integration of IoT and AI into the architecture of digital twins is the keystone for realizing their full potential. It harnesses the relentless stream of data from our increasingly connected world, combined with the unparalleled analytical capabilities of AI, to create intelligent replicas that are as dynamic as the real-world systems they emulate. It's not just about keeping pace with change—it's about anticipating and driving it. As you explore the exciting landscape of digital twins, consider how you might harness these integrations to propel your organization or community to new heights of innovation and resilience.

Chapter 3: Applications in Manufacturing

Manufacturing has always been the backbone of industrial progress, and with digital twins, this sector is poised to reach unprecedented heights of efficiency and innovation. Imagine a factory where real-time data and simulations create a seamless bridge between the physical and digital realms, providing insights that are as dynamic as the production lines themselves. In this transformative landscape, digital twins allow manufacturers to fine-tune processes, predict failures before they occur, and adapt swiftly to changes, all while maintaining optimal resource use and minimizing downtime. Through enhanced production efficiency and real-time monitoring systems, digital twins enable a level of precision and responsiveness that traditional manufacturing methods simply can't match. As the manufacturing industry continues to embrace these cutting-edge technologies, the potential for operational excellence and sustainable practices promises to shape a more resilient and agile future. This isn't just evolution; it's a revolution waiting to set new standards across the global manufacturing landscape.

Enhancing Production Efficiency

In the heart of manufacturing, where precision and performance are paramount, digital twins stand as transformative allies. By blending the physical and virtual realms, these advanced simulations provide a comprehensive view of production processes in unprecedented detail.

With digital twins, production efficiency isn't just enhanced; it's redefined.

Manufacturers have long sought ways to optimize their operations, reduce waste, and increase output without sacrificing quality. The introduction of digital twins into the manufacturing process reshapes these ambitions into reality. With a digital replica of production lines, manufacturers are empowered to simulate and analyze processes before executing them in real life. This foresight allows them to test, tweak, and troubleshoot without the downtime or resource expenditure traditionally needed for such exhaustive testing.

Consider, for instance, the capacity of digital twins to model entire assembly lines. By creating a virtual counterpart, manufacturers can predict bottlenecks or inefficiencies. This not only illuminates the complexities of production but also promotes smarter decisions and sharper allocation of resources. Imagine being able to tweak a single variable in a virtual setting, observe the resultant effects, and adjust accordingly—all before making tangible changes. It ushers in a more agile and responsive manufacturing environment.

Moreover, digital twins facilitate robust process optimization. By collecting data from IoT-enabled machines and integrating it with AI predictive models, production lines become more intelligent and self-regulating. These smart factories adapt and learn, continuously optimizing themselves with each cycle of the production process. Not only does this improve efficiency, but it also minimizes interruptions, reduces energy consumption, and enhances product quality.

The predictive power of digital twins extends beyond optimization, offering manufacturers a proactive stance in maintenance. Where traditional maintenance relied on scheduled checks or reactive repairs, digital twins allow for predictive maintenance. Leveraging real-time data, these simulations can predict equipment failures before they occur, guiding timely interventions that avert costly downtimes and extend machinery life.

Collaboration also takes on new dimensions with digital twins. Engineers, designers, and operators can work in tandem from disparate locations, interacting with the digital model to brainstorm, test, and refine ideas. The visual and interactive nature of digital twins democratizes information and insights, driving collective innovation and fostering a culture of continuous improvement.

Yet, the transformative impact of digital twins on production efficiency isn't just about the technology itself, but how it empowers people. It shifts the focus from traditional labor-intensive processes to more strategic and analytical roles. Workers can now focus on value-added tasks while digital twins handle the complexities of monitoring and optimization.

Furthermore, educational implications abound. The use of digital twins enables training in virtual environments, offering a safe space for workers to learn and experiment without fearing real-world consequences. This enhances the skillsets of workers, making the workforce more adaptable and ready for shifts in production needs.

The strategic deployment of digital twins also helps in demand forecasting. By simulating different production scenarios, businesses can better align their outputs with market demands. This elasticity in production helps companies adapt rapidly to changes, reducing lead times and ensuring that customer needs are met promptly and efficiently.

Of course, as we embrace this digital revolution, it's crucial to remain aware of potential challenges. Implementing digital twins requires substantial initial investment in technology and training. There's also the matter of data management; ensuring that systems are in place to capture, store, and analyze the massive amounts of data generated is critical. However, when these hurdles are addressed, the returns in efficiency and productivity are significantly rewarding.

In conclusion, the enhancement of production efficiency through digital twins is not merely a tale of advanced technology but a narrative of meaningful transformation. By marrying the precision of digital simulations with real-world processes, manufacturers are crafting pathways to innovation that were once deemed unattainable. As the landscape of manufacturing continues to evolve, the role of digital twins in driving efficiency remains unwavering, setting a new paradigm for what's possible.

Real-Time Monitoring Systems

In the fast-paced realm of modern manufacturing, real-time monitoring systems powered by digital twins are revolutionizing the way industries operate. These systems are not just technological novelties—they're transformative tools that can reshape entire manufacturing processes. With the integration of digital twins, manufacturers can stay ahead of the curve by continuously monitoring equipment, operations, and environments, leading to a new era of efficiency and responsiveness.

At the heart of these systems lies a simple yet powerful concept: by creating a digital replica of a physical asset, manufacturers can monitor the real-world performance of their operations in real-time. This isn't merely about data collection; it's about gaining insights that drive smarter decision-making. For instance, digital twins track machine performance, energy consumption, and product quality, providing a comprehensive view that empowers operators to make adjustments on-the-fly. In essence, they are a window into the soul of the manufacturing process, revealing details that were previously hidden or inaccessible.

One of the key advantages of implementing real-time monitoring systems is the enhancement of predictive maintenance. Traditional maintenance schedules often rely on fixed intervals or reactive troubleshooting, which can lead to downtime and inefficiencies. In contrast, digital twins enable condition-based maintenance strategies. By continuously analyzing data, these systems can predict when a machine component may start to fail, thus allowing for proactive interventions. This shift not only minimizes downtime but also extends the lifespan of equipment, which can mean significant cost savings over time.

Moreover, digital twins facilitate the synchronization of production lines through real-time monitoring. In complex manufacturing environments, operations must be meticulously coordinated to avoid bottlenecks and delays. With digital twins, different parts of the production process can communicate seamlessly, ensuring smooth transitions and reduced idle times. This is particularly critical in just-in-time manufacturing scenarios, where timing is everything. The real-time data feedback loops created by digital twins help maintain optimal flow and reduce waste, embodying lean manufacturing principles.

Another vital application of real-time monitoring systems is quality control. In manufacturing, quality is non-negotiable, and digital twins play a critical role in ensuring that products meet the highest standards. By monitoring every aspect of the production process, from raw materials to final assembly, real-time systems can quickly identify deviations or defects. Manufacturers can then address issues promptly, adjusting processes to correct errors before products are completed. This immediate feedback loop not only improves product quality but also reduces waste and increases customer satisfaction.

The integration of digital twins in real-time monitoring also facilitates increased flexibility in manufacturing. In a world where consumer preferences and market demands change rapidly, flexibility is a critical asset. Digital twins allow manufacturers to simulate changes in product design or production processes virtually before implementing them physically. This capability enables manufacturers to adapt quickly to new requirements, alter production runs, and innovate without the risk of costly errors.

In addition to process improvements, digital twins enhance the decision-making processes within manufacturing operations. Managers and decision-makers benefit from comprehensive dashboards that aggregate and analyze data in real-time. These dashboards provide an overarching view of performance metrics, allowing for quick analysis and rapid response to challenges. This accessibility to actionable insights empowers leaders to make informed decisions that align with business goals while promoting a culture of continuous improvement.

Real-time monitoring systems supported by digital twins also foster collaboration across different levels of an organization. Transparency in operations data ensures that teams from the shop floor to the boardroom are working with the same information, promoting coordinated efforts and aligning objectives. This collaborative approach breaks down data silos and encourages a more integrated workflow, where every stakeholder is informed and engaged in the pursuit of operational excellence.

The benefits of real-time monitoring systems extend beyond individual manufacturing facilities. Companies with multiple sites can use digital twins to achieve consistency and standardization across locations. Real-time data enables centralized management teams to oversee operations, identify best practices, and implement improvements on a global scale. This consistency streamlines processes and reduces variability, allowing for uniform product quality and operational efficiency, regardless of location.

Of course, the implementation of these systems is not without its challenges. The integration of real-time monitoring systems requires substantial investment in technology, training, and infrastructure. Companies must address data privacy concerns and ensure robust cybersecurity measures to protect sensitive information. However, the potential benefits far outweigh these challenges, offering a meaningful return on investment for forward-thinking manufacturers.

In conclusion, real-time monitoring systems in manufacturing, powered by digital twins, represent an exciting frontier in industrial innovation. They provide manufacturers with the tools needed to navigate the complexities of contemporary production environments, driving efficiency, quality, and agility. As technology continues to evolve, these systems will only become more sophisticated, further strengthening their role in shaping the future of manufacturing. Embracing this digital transformation will not only enhance operational performance but also ensure that manufacturers remain competitive in an increasingly dynamic global market.

Chapter 4: Transforming Healthcare with Digital Twins

The healthcare sector stands on the cusp of a profound transformation, driven by the integration of digital twins into its core processes. As digital replicas of physical entities, these twin models offer unprecedented opportunities for enhancing patient care, optimizing resource allocation, and revolutionizing medical research. By enabling real-time monitoring and predictive analytics, digital twins support a shift towards patient-centric innovations that personalize treatment plans and improve health outcomes. In research arenas, the ability to simulate complex biological systems accelerates the discovery of novel therapies and medical procedures, making breakthroughs more achievable than ever before. With the dynamic interplay between technology and healthcare, digital twins are not just a tool but a catalyst for systemic change, inspiring healthcare professionals to envision a future where technology and care are seamlessly intertwined. This chapter explores how digital twins are shaping new frontiers in healthcare, setting the stage for a paradigm shift that redefines the very essence of medical practice and research.

Patient-Centric Innovations

The rise of digital twins in healthcare is more than just a technological breakthrough; it's a paradigm shift towards prioritizing patient-centric care. At its core, patient-centric innovations aim to tailor healthcare experiences, ensuring that each patient receives personalized treatment plans based on their unique physiological and medical data, rather than standardized protocols. This approach revolutionizes the way healthcare providers interact with patients, placing the individual at the center of care strategies.

Digital twins offer an unprecedented level of detail in understanding a patient's condition. By creating a virtual representation of an individual patient's anatomy and health data, healthcare practitioners can explore multiple treatment scenarios without risk. This not only enhances predictive capabilities but also reduces the trial-and-error aspects of traditional treatment approaches. With digital twins, doctors can simulate the impact of different medications, lifestyle changes, or surgical procedures before making any physical interventions. This level of personalized simulation was unthinkable just a few years ago.

The application of digital twins in chronic disease management exemplifies the patient-centric model. Many chronic conditions, like diabetes and heart disease, demand continuous monitoring and frequent adjustments in treatment. Digital twins can integrate real-time data from wearable devices, electronic health records, and even genetic information to provide ongoing insights into a patient's health. As a result, care can be dynamically adjusted to better suit individual needs, significantly improving outcomes and enhancing quality of life. For patients, this means fewer hospital visits, personalized care plans, and a proactive rather than reactive approach to managing their health.

Take, for example, the management of diabetes. A digital twin of a diabetic patient could continuously monitor glucose levels, lifestyle habits, and other health parameters. In response to deviations from the norm, the digital twin can analyze potential causes and recommend adjustments before complications arise. This level of precise management helps to prevent crises and maintain balanced blood sugar levels, ultimately reducing long-term health risks for the patient.

Furthermore, digital twins facilitate superior communication between patients and healthcare providers. By using a visual and interactive representation of health data, patients are empowered to understand their conditions better. This enhanced understanding fosters collaborative decision-making, encouraging patients to take an active role in their health management. The shift to a more informed patient base transforms the doctor-patient relationship into a partnership, promoting adherence to treatment plans and improving health outcomes.

Pediatric care has also reaped benefits from digital twin technology. Children's healthcare requires special considerations due to their ongoing development. Digital twins allow for the careful monitoring of growth patterns, ensuring that interventions are appropriately timed. For instance, in managing congenital heart defects, digital twins can simulate how a child's heart will respond to growth over time, guiding surgical decisions and interventions to optimize future health.

In the realm of predictive healthcare, digital twins are game-changers. By modeling potential future health scenarios, they provide insights that help foresee diseases even before symptoms emerge. This predictive quality extends to genetic disorders and conditions influenced by environmental factors. Knowing a patient's predisposition and environmental interactions allows for early interventions, such as lifestyle modifications or preventive treatments, effectively curing ailments before they manifest.

The integration of artificial intelligence with digital twins takes patient-centric care to another level. AI algorithms process vast arrays of data far beyond human capabilities, identifying patterns and correlations for more refined insights. When combined with a digital twin, AI can predict disease progression and responses to treatments with remarkable accuracy. This merging of technologies ensures that every patient receives a truly tailored healthcare experience, continuously refined as technology advances.

Admittedly, the journey towards widespread adoption of digital twins in healthcare faces hurdles. Data privacy and integration challenges persist, with concerns over how sensitive health information is gathered, stored, and utilized. However, as promising as they are, these obstacles are not insurmountable. The value that digital twins provide in aligning healthcare to individual needs motivates continual advancements in data security and interoperability within healthcare systems.

Digital twins are also enhancing the design and execution of clinical trials. By simulating patient groups through virtual models, researchers can identify candidate interventions more swiftly and accurately. This accelerates the discovery of new therapies and reduces the trial period, lowering costs and expanding access to innovative treatments. Patient-generated data leads to more inclusive and comprehensive trials, accommodating variability in real-world populations.

From a broader perspective, what digital twins offer is a realignment of healthcare priorities, focusing less on symptoms and more on holistic, personalized well-being. Health systems that adopt this model become more proactive, guiding patients toward healthier lifestyles, preventing illness rather than reacting to it, and reducing the overall burden on healthcare facilities. Ultimately, this reflects an evolution towards a healthcare ecosystem that is preventive, personalized, participatory, and predictive.

In conclusion, digital twins stand as a beacon of hope for achieving a new era of patient-centric care. They are transforming the practice of medicine into a science of precision and prediction, where each individual's health journey is uniquely catered to. As digital twin technology matures and integrates more deeply into healthcare systems, the possibilities for patient-centric innovations are endless, holding the promise to redefine what it means to receive care in the twenty-first century.

Revolutionizing Medical Research

In a world where technology is advancing at an unprecedented pace, digital twins have emerged at the forefront of innovation in medical research. Their ability to simulate real-world phenomena in a virtual environment is a game-changer, enabling breakthroughs that were once thought to be beyond our reach. Rather than working with a traditional trial-and-error approach, researchers now utilize digital twins to experiment with complex biological systems safely and efficiently.

Digital twins bridge the gap between theoretical models and biological reality. This capability is transforming how medical research is conducted, propelling us into an era where personalized and precision medicine becomes increasingly accessible. For decades, understanding the intricate workings of the human body presented a monumental challenge. However, with digital twins, researchers can now create highly detailed virtual models of organs or even individual cells. This granularity allows for precise simulations and testing, which in turn enhances the accuracy of research outcomes.

Beyond modeling isolated biological components, digital twins enable the integration of data from heterogeneous sources, creating comprehensive simulations of entire biological systems. This holistic view allows for a better understanding of the interplay between various biological factors, such as genetics, environment, and lifestyle, that influence health and disease. Researchers can unravel the complexities of multifactorial diseases with unprecedented precision, opening the door to new treatment paradigms that are tailored to individual patients.

Perhaps most strikingly, digital twins empower researchers to conduct virtual clinical trials. Traditional clinical trials are often fraught with significant time and cost constraints, not to mention ethical considerations. Digital twin technology mitigates these issues by allowing researchers to simulate trials in a virtual environment before or alongside real-world implementation. The ability to model different patient responses to treatments—even before the first real-world application—streamlines the drug development process, potentially bringing life-saving therapies to market faster and more safely.

Moreover, digital twins facilitate iterative testing and hypothesis refinement in a risk-free setting. Researchers can manipulate variables, observe potential outcomes, and refine hypotheses with minimal risk and maximum insight. This iterative process accelerates innovation and enhances the reliability of conclusions drawn from research, vastly improving the quality of scientific inquiry.

Interdisciplinary collaboration finds fertile ground in the context of digital twins. The integration of knowledge from various domains—such as biology, computer science, data analytics, and engineering—fosters a collaborative environment where diverse teams work towards a common goal. The insights gleaned from one field of study can complement and enhance those from another, creating a synergistic effect that significantly amplifies the impact of research initiatives.

In the journey towards using digital twins in medical research, challenges remain. High-quality data acquisition and management are critical, as the fidelity of digital twin models largely depends on the accuracy and breadth of data they are fed. Data privacy and security issues also loom large, necessitating robust safeguards to protect sensitive patient information. Yet, these challenges present opportunities for technological advancements in areas like data encryption, anonymization, and distributed ledger technologies to enhance trust and security.

As digital twin technology continues to evolve, its adoption in medical research is bound to inspire a paradigm shift that goes beyond academia into the clinical world. Imagine a future where every medical professional has access to a personalized digital twin of their patient—an exact virtual replica that can help anticipate disease progression, optimize treatment plans, and predict outcomes. Not only would this revolutionize patient care, but it would also transform the role of healthcare professionals, empowering them with tools for unprecedented precision and foresight.

The motivation behind integrating digital twins into medical research is clear: the potential for discovery, innovation, and improved healthcare outcomes is immense. The narrative of digital twins in medical research is one of hope—a promise for a brighter future where diseases that have long eluded our grasp may finally succumb to our scientific prowess.

Ultimately, digital twins offer the vision of a healthcare landscape that is not just reactive but predictive, not just general but personalized. By harnessing the power of digital twins, researchers stand on the cusp of breakthroughs that will redefine what is possible in medicine, emphasizing a future where the boundaries of medical research continue to expand and redefine the destinies of countless individuals.

Chapter 5: Digital Twins in Smart Cities

Digital twins hold the transformative potential to redefine how we conceive and manage urban spaces. As cities become smarter, they grapple with complex challenges—resource optimization, environmental sustainability, and enhanced public services. Digital twins offer a compelling solution by creating dynamic digital counterparts of physical city environments. These virtual models empower city planners and decision-makers to simulate and analyze the impacts of proposed developments before they become reality, preventing costly mistakes. They can enhance city infrastructure efficiency by optimizing traffic flows and energy usage, contributing to sustainability goals. Bridging the physical and digital realms, digital twins enable real-time responses to changing urban needs, fostering a city ecosystem that is not just reactive but also predictive. They are catalysts for innovation, making our urban landscapes more livable, responsive, and sustainable, ultimately embracing the future that once seemed distant into the sphere of current possibilities.

Urban Planning and Development

Digital twins are steadily revolutionizing urban planning and development, allowing cities to evolve in smarter, more efficient ways. By creating detailed, virtual models of physical spaces, urban planners can simulate and evaluate different scenarios, enabling more informed decision-making. This innovative approach enhances the precision and efficiency of urban planning, ensuring that resources are used wisely and projects meet community needs effectively.

One of the most significant advantages of digital twins in urban planning lies in their ability to integrate vast amounts of data. Cities generate enormous datasets through various sources such as traffic systems, environmental sensors, and social media. By using digital twins, these datasets can be aggregated into a single, coherent model. This model provides a dynamic platform where planners can visualize and interact with the city's current and future states. It aids in identifying patterns and predicting outcomes, thus facilitating decisions that support sustainable development.

Urban centers face constant challenges due to population growth and infrastructure demands. Digital twins help mitigate these challenges by allowing planners to test various development scenarios without the cost or risk of physical trials. For example, transport systems can be modeled to predict traffic flow improvements or the impact of new public transit routes. By simulating these changes, cities can optimize their infrastructure investments and improve citizen experiences.

Environmental sustainability is another key area where digital twins provide immense benefits. Urban planners can model the effects of green initiatives, such as the increase of green spaces or the implementation of renewable energy sources. These simulations can help in understanding how changes affect the urban ecosystem, assisting planners in making choices that reduce carbon footprints and enhance air quality.

Moreover, digital twins empower collaborative planning efforts among various stakeholders. Through these platforms, architects, engineers, city officials, and community members can seamlessly share insights and align their objectives. This cooperation leads to well-integrated solutions that fulfill the unique requirements of different community sectors, ensuring that development initiatives are both effective and inclusive. Enhanced transparency in project development fosters public trust and paves the way for smoother implementation.

The use of digital twins enables predictive maintenance and efficient resource allocation. By continuously analyzing real-time data from infrastructure sensors, potential system failures can be identified before they occur, reducing downtime and repair costs. This proactive approach ensures the longevity of urban assets, resulting in cost savings and enhanced service reliability.

In storm water management, digital twins play a crucial role by simulating drainage systems and modeling weather patterns. Planners gain insights into potential flood risks and can design robust infrastructure to mitigate these threats. This capability is especially critical in areas prone to extreme weather events, where effective management of storm water systems can save lives and property.

In addition, digital twins can dynamically model socio-economic trends. By integrating demographic and economic data, planners can forecast growth patterns, housing demands, and economic developments. This information is vital for making informed decisions on zoning regulations, public facilities, and business incentives that drive economic vitality.

The implementation of smart technologies, such as IoT, further enriches digital twin models by providing a continual flow of data. This interconnectivity creates a real-time feedback loop that improves urban responsiveness. For instance, smart streetlights equipped with environmental sensors can adjust to weather conditions or traffic patterns, enhancing energy efficiency and public safety.

Even cultural and historical elements of urban landscapes can be preserved and enhanced through digital twins. Virtual models allow planners to visualize how new structures will coexist with historical sites, ensuring that development honors the city's heritage while meeting contemporary needs. By considering aesthetic and cultural impacts, digital twins help balance modernization with preservation.

Ultimately, the integration of digital twins in urban planning fosters a culture of innovation. By providing a sandbox for experimentation, cities can explore ambitious ideas with confidence, unburdened by the physical and financial constraints of traditional prototyping. This innovative mindset not only accelerates the development process but also inspires planners to rethink conventional approaches, leading to creative solutions for modern urban challenges.

Digital twins are more than just tools; they're transformative forces driving the future of urban development. As cities continue to grow and the complexity of urban environments increases, the role of digital twins will become even more critical. They offer a pathway to smarter, more livable cities, where data-driven insights guide decisions and enhance quality of life for all residents. With digital twins, the cities of tomorrow will not only meet the demands of today but will do so in ways that are sustainable, equitable, and innovative.

Improving Public Services and Infrastructure

In an age where urban environments are growing at an unprecedented pace, the complexity of managing public services and infrastructure poses significant challenges. However, digital twins are reshaping our possibilities. By creating dynamic, data-driven simulations of real-world objects and systems, cities can leverage digital twins to revolutionize how public services are delivered and infrastructure is managed. This innovative technology promises not just incremental improvements but transformative changes that aim to uplift urban life quality.

Imagine a city where traffic congestion is minimized because traffic signals dynamically adjust in real-time based on actual traffic conditions. With digital twins, this is a reality. They enable city planners to simulate various conditions and traffic scenarios, leading to a more efficient flow of vehicles. These simulations help in uncovering the ripple effects of new traffic patterns, allowing for evidence-based modifications that are critical to improving urban mobility.

Moreover, digital twins play a crucial role in optimizing public transportation networks. They provide a virtual model where urban planners can test changes without disrupting existing services. By analyzing data collected from sensors in trains, buses, and stations, digital twins assist in predicting maintenance needs and improving service schedules. This ensures that public transport systems are reliable and can adapt swiftly to the evolving demands of the city's residents.

Another profound impact of digital twins is in water and waste management systems. Traditional systems struggle with inefficiencies that lead to the wastage of resources and potential hazards to public health. Digital twins revolutionize this by providing precise models of water usage, waste generation, and treatment facilities. City managers can use these virtual models to optimize the allocation of resources, reduce waste, and ensure the sustainability of these vital services.

Energy management is another cornerstone of city infrastructure where digital twins are making strides. By integrating renewable energy sources into city grids, digital twins allow for the simulation of different energy consumption patterns and supply scenarios. This helps in balancing supply and demand while minimizing wastage. As cities grapple with their energy footprints, such dynamic modeling ensures that sustainability goals are met efficiently.

Infrastructure maintenance, traditionally a reactive process often delayed until failures occur, is now entering a proactive era. Digital twins allow for predictive maintenance models that anticipate when parts of a city's infrastructure might fail, from bridges and roads to sewage systems. By simulating different stress factors and degradation patterns, it becomes possible to perform timely interventions that save cities from costly repairs and potential hazards.

Emergency response services also find an invaluable ally in digital twins. They provide simulation environments where different disaster scenarios such as floods, fires, or earthquakes can be modeled. These models allow emergency services to strategize effectively, improving response times and decision-making processes during actual events. Training scenarios can be run on the digital twins, providing real-world expertise without real-world risks.

Additionally, digital twins contribute to better urban planning. They offer planners a chance to visualize the impact of new developments or zoning changes in a simulated environment. With this capability, urban planners can foresee issues such as overpopulation, underdeveloped areas, or potential environmental impacts. This foresight ensures that urban development is both sustainable and resilient against future uncertainties.

The adoption of digital twins also enhances citizen engagement. By making the data accessible and understandable, residents can see how simulated changes might impact their daily lives. This transparency fosters community involvement and ensures that the development path aligns with resident needs and desires. When citizens understand the implications of new infrastructure projects, they are more likely to support and participate in public initiatives.

Moreover, by forging collaborative ecosystems between government agencies, private sectors, and academia, digital twins encourage innovation and resource sharing. Cities can tap into a wealth of expertise and data through these partnerships, thereby accelerating the evolution and application of digital twin technology in urban settings. Such collaboration leads to the creation of comprehensive smart city solutions that are robust and scalable.

The potential of digital twins in refining public services and infrastructure is immense, but it requires a shift in perspective. Digital twins should not be seen merely as tools for troubleshooting existing problems, but rather as strategic enablers for holistic urban innovation. They challenge us to rethink how cities operate and interact with their inhabitants. Thus, embracing digital twins unfolds a future replete with possibilities that promise not just smarter cities, but healthier, safer, and more connected urban experiences.

Chapter 6: Automotive Industry Innovations

The automotive industry is shifting gears, accelerating toward a future where digital twins play a pivotal role in revolutionizing vehicle performance and safety. By adopting these advanced virtual models, manufacturers can simulate real-world conditions, enhancing design, testing, and production processes with unprecedented precision and speed. This transformative power extends beyond manufacturing, driving innovations in motorsport with virtual twins enabling teams to fine-tune strategies and vehicles for optimal performance under varying conditions. The impact is evident: a seamless merger of real and digital realms promises not only enhanced safety and performance but also paves the way for more sustainable and efficient transportation solutions. As the industry navigates this digital highway, it's crucial for stakeholders to embrace these technological advancements, propelling us toward a smarter and safer automotive future.

Vehicle Performance and Safety

The evolution of vehicle performance and safety is at the heart of automotive industry innovations. The infusion of digital twins into this arena is nothing short of revolutionary. These virtual replicas allow engineers to simulate, track, and analyze a vehicle's lifecycle with extraordinary precision. What's fascinating is how digital twins can enhance every pivotal aspect of a vehicle's journey, from the initial design to real-world performance and, ultimately, end-of-life recycling.

First, let's delve into vehicle performance. Performance optimization always stood as a complex challenge given the myriad of variables involved. Factors such as aerodynamics, engine efficiency, and structural integrity are traditionally tested through time-consuming iterations in physical spaces. However, digital twins streamline this entire process. By creating a virtual model, engineers can experiment with countless configurations and swiftly identify optimal designs long before they build a physical prototype. This not only accelerates development but ensures each vehicle rolls off the line performing at its peak.

In terms of safety, digital twins make a significant impact by simulating real-world conditions and potential hazards. They allow automakers to test scenarios that would be either dangerous or impractical to conduct in the physical world. For instance, analyzing crash impacts, weather effects, and unexpected driver behaviors can be assessed with startling accuracy in a virtual environment. This ability to predict and preemptively address safety concerns has tremendous implications for reducing accidents and saving lives.

Moreover, digital twins enhance vehicle performance by facilitating continuous monitoring and predictive maintenance. Real-time data from sensors embedded within a vehicle feed into its digital twin, allowing for constant performance assessment. This data-driven approach enables the early detection of issues, preventing breakdowns before they occur. In a world where downtime equates to financial loss, this advancement helps industries maintain productivity and efficiency.

Beyond internal mechanics, digital twins play a critical role in shaping the external attributes of a vehicle, such as aerodynamics. A car's interaction with air involves a delicate balance that impacts fuel efficiency, speed, and handling. Engineers can use digital twins to simulate airflow over the vehicle's surface, identifying spots that generate drag and adjusting designs to minimize it. These optimizations lead to vehicles that are not only faster and more efficient but also environmentally friendly due to reduced emissions.

From a safety perspective, digital twins provide an unparalleled opportunity for automakers to collaborate with urban planners. As smart cities evolve, there is a growing need to consider how vehicles interact with their environments. Digital twins can model traffic patterns, pedestrian safety, and emergency response scenarios, ensuring that vehicles contribute to, rather than detract from, the sustainability and safety of urban life.

Furthermore, digital twins empower manufacturers to incorporate advanced driver-assistance systems (ADAS) with greater precision. Components like adaptive cruise control, lane-keeping systems, and automated braking systems can be tested and validated in the virtual realm. This precursory validation is instrumental in reducing costly recalls and enhancing consumer trust by delivering reliable, safe vehicles.

Another significant advantage lies in the simulation of extreme conditions, which might not otherwise be feasible or safe to test physically. For example, digital twins can replicate icy roads, high-speed maneuvers, or collision scenarios involving multiple vehicles. Evaluating vehicle performance and occupant safety in these extreme environments allows manufacturers to develop robust solutions that ensure resilience and reliability.

As performance and safety standards continue to evolve, so too do regulatory frameworks. Digital twins offer a vital tool for navigating these complex regulatory waters. Automakers can demonstrate compliance through simulations and data-driven evidence, providing regulators with transparent and comprehensive assurance of safety and performance standards. This proactive approach can streamline the homologation process and speed up time-to-market.

Digital twins aren't just about technology—they're about transformation. They catalyze a shift in how the automotive industry perceives innovation, bridging traditional craftsmanship with cutting-edge technology to redefine what vehicles can achieve. This paradigm shift ultimately inspires a vision of a future where safety is paramount, performance is optimal, and the entire lifecycle of vehicles contributes to sustainable industry practices.

In the realm of vehicle safety, the implications of digital twins extend beyond the confines of Hollywood-style crash tests. Through analytic prowess, manufacturers can explore phenomena such as material fatigue and long-term stress factors that impact vehicle integrity over time. By understanding how cars age and wear, engineers can design for durability and reliability, extending the vehicle's lifespan and value.

Digitization via digital twins ensures that vehicle performance and safety continue to advance in tandem. These virtual replicas are not static; they evolve with the vehicle, gathering data and improving with every journey on the road. Such feedback loops close the gap between expected and actual performance, enabling the industry to fine-tune products dynamically and continuously.

In conclusion, the integration of digital twins within the realm of vehicle performance and safety signifies not merely an incremental improvement but a transformative leap. Through these virtual models, the automotive industry achieves a new height of innovation, ensuring vehicles that are not just machines of movement but paragons of safety and performance. It's an inspiring journey, reshaping the automotive landscape for technologies yet to come and underscoring the importance of embracing digital twins in the relentless pursuit of excellence.

Digital Twins in Motorsport

Motorsport sits at a unique intersection of high-speed thrills and cutting-edge technology. This realm isn't just about blistering velocities and daring drivers; it's an arena where technology evolves rapidly, and the stakes are nothing short of victory and glory. One of the most fascinating advancements in recent years has been the application of digital twins in motorsport. These virtual replicas are transforming how teams innovate, strategize, and perform on the track.

In essence, digital twins in motorsport offer a detailed virtual model of the vehicle, driver, and the environmental conditions they face. They simulate complex scenarios, perform rigorous tests virtually, and provide insights that are pivotal for the real-world interpretations of these complex systems. For teams aiming to fine-tune every aspect of their performance, these digital comparisons enable unprecedented precision.

Consider a Formula 1 car, which represents the pinnacle of automotive engineering and performance. The integration of digital twins allows teams to simulate endless combinations of track conditions, weather variables, and mechanical adjustments before the tires even touch the track. With the unpredictability that racing brings—like sudden rain, temperature shifts, or a change in wind direction—having a virtual model to test responses in advance is a game-changer.

Moreover, digital twins enhance data management and analysis. Motorsport engineers collect enormous data sets during races or testing sessions—everything from tire temperatures to aerodynamic performance. Integrating these data streams with digital twins, teams can create detailed simulations that provide insights into how a car might behave under specific conditions. This leads to informed decision-making, helping develop strategies that weren't feasible in the past.

Driver preparation has also embraced digital twins. Many racing teams use them to build hyper-realistic simulators that replicate real track conditions and car dynamics. This kind of immersive training isn't just about preparing for a race; it's about enabling drivers to experience potential scenarios and responses in a controlled environment. The result? Drivers who adapt quickly and make decisions faster than ever when they encounter similar situations on the track.

Leaning into the realm of predictive analysis, digital twins bring a strategic edge. Imagine predicting the mechanical wear and tear on a car component before it leads to a catastrophic failure. With digital twins, teams use data not only from the vehicle but from historical performance patterns, understanding when and where failures might arise. This predictive capability translates into strategic advantages by optimizing pit stops, extending the life of components, and significantly reducing the risk of mid-race technical failure.

As sustainability becomes a cornerstone in every industry, motorsport isn't immune to the pressures to reduce its environmental impact. Digital twins facilitate this by minimizing the need for physical prototypes and extensive on-track testing, both of which require energy and resources. Instead, teams can conduct virtual tests, sparing resources by solving engineering puzzles in the digital realm, which not only drives efficiency but also reduces environmental footprints.

Beyond the track, fans engage with the sport through these innovations. Motorsport enthusiasts crave more than just spectating—they want insights and interactivity. Digital twins can provide live, interactive experiences where fans view race dynamics in real-time through digital models. This transparency not only heightens the viewing experience but also builds a more profound connection with the sport, reinforcing fan loyalty.

The ripple effect of digital twin technology extends to motorsport engineering itself. By refining designs swiftly in the virtual world, concepts that might have remained impractical or cost-prohibitive now come to life. This creative freedom fosters a new wave of innovation, propelling motorsport teams to blaze trails and redefine what's possible within the boundaries of automotive racing.

While the promise of digital twins in motorsport is bright, their adoption isn't without challenges. Integrating new technologies into teams' existing structures requires vision and change management. Resources like talent, time, and budget must be thoughtfully allocated to ensure seamless integration of these digital methodologies. Teams must navigate this landscape with agility and adaptability, being willing to learn quick lessons from approaches that might fail to meet expectations initially.

Ultimately, digital twins are revolutionizing motorsport by melding the physical with the digital in ways that seem almost magical. They drive efficiencies and innovational leaps, pushing the limits of what humans and their machines can achieve on racetracks across the globe. As we peer into the future, the bond between motorsport and technological evolution will continually redefine the contours of this exhilarating sport, painting a world where data drives decision-making and digital models epitomize perfection in performance.

From the roar of engines to the heartbeat metronome of precisely tuned cars, the marriage of speed and technology is happening faster than one can blink in motorsport. Digital twins are not just participants in this journey; they're architects, laying out new paths on which to race forward.

Chapter 7: Energy Sector Transformation

The energy sector stands on the precipice of a profound makeover, with digital twins playing a pivotal role in driving this change. Imagine a world where power plants operate at peak efficiency, thanks to real-time data and predictive analytics that help anticipate equipment failures before they occur. Digital twins enable operators to model scenarios for smart grids, dynamically shifting demand and supply with renewable sources like solar and wind. This technology doesn't just optimize current operations; it transforms them by providing insights that lead to a cleaner, more sustainable energy landscape. The implications are enormous—from reducing carbon footprints to lowering costs for consumers. For industry professionals and decision-makers, embracing digital twins unlocks unprecedented possibilities, leading the way in energy sustainability and innovation. As we move forward, these virtual counterparts are essential in turning challenges into opportunities, propelling the sector toward a more resilient and efficient future.

Optimizing Power Plants

The energy sector stands on the cusp of a digital revolution, with digital twins poised to optimize power plants like never before. Picture a virtual replica of a power plant, operating in real-time, mirroring the physical plant's every performance metric and predictive model. This

Intelligent mirror-image is a digital twin, brimming with potential to transform how we generate, distribute, and consume energy.

Let's dive into how digital twins are shaping the future of power plant management. The principal lure of adopting digital twins in power plants is their ability to enhance operational efficiency and minimize downtime. Through continuous monitoring and predictive analytics, digital twins can forecast equipment failures and suggest preventive maintenance measures before a real problem occurs. This predictive maintenance, grounded in data, reduces unexpected shutdowns that can cost millions.

Now, imagine the complex ecosystem of a power plant. Numerous components, from turbines to transformers, need seamless synchronization. Digital twins offer a comprehensive, real-time view of these interconnected systems, helping operators visualize the entire plant's operational health at a glance. They allow for scenario testing and optimization of processes without risking any physical damage or interruption. Operators can run simulations to determine the most efficient settings, saving energy and cost.

Moreover, digital twins foster an adaptable and dynamic response to energy demand fluctuations. They enhance control over power plant operations, enabling the modulation of output to meet real-time demand with optimal energy consumption. This ability significantly contributes to reducing energy waste, a crucial step towards achieving a sustainable energy future.

Flexibility is another vital advantage. The integration of renewable energy sources such as wind and solar into power plants is crucial for the future. However, the variability of these sources poses significant challenges. Digital twins assist in modeling these fluctuations, providing insights into how to integrate renewables sustainably with minimal impact on power reliability. By analyzing historical data and forecasting upcoming weather patterns, digital twins help manage the balance between renewable and traditional energy resources efficiently.

Transparency and decision-making are radically improved too. With visual dashboards and real-time data analysis, plant operators and decision-makers gain deeper insights into every facet of operations. This transparency leads to more informed decisions, fostering a culture of continuous improvement. Data-driven insights help identify opportunities for efficiency improvements, emission reductions, and performance enhancements.

The use of digital twins in optimizing power plants isn't just about operational benefits. They hold promise for sustainability goals. By optimizing processes and reducing waste, digital twins contribute to lowering carbon emissions. Detailed simulations can illustrate potential emission reductions and highlight areas where the plant can operate more sustainably.

One can't overlook the contribution of artificial intelligence in enhancing digital twins' capabilities. Machine learning algorithms analyze vast datasets to uncover patterns and insights that would be otherwise impossible or impractical for humans to discern. As these algorithms evolve, they offer increasingly sophisticated predictions and optimizations, continually refining their models based on new data.

Collaborations and partnerships in the technology and energy sectors play a vital role in digital twin adoption. By leveraging collective expertise, stakeholders can overcome implementation challenges, establish standards, and innovate more rapidly. Power plant operators, technology providers, and academic institutions working together can develop customized solutions for specific challenges, ensuring digital twins meet diverse operational needs.

However, the journey to deploying digital twins in power plants isn't without challenges. Data security and integration across various platforms and technologies require robust solutions. Stakeholders must prioritize cybersecurity protocols to protect sensitive data from breaches. Developing interoperable platforms that can easily integrate with existing systems is also crucial for seamless operation.

Despite these challenges, the momentum towards adopting digital twins is undeniable. Power plants across the globe are beginning to embrace this technology, driven by the promise of unparalleled operational efficiency and sustainability. As the pressure mounts to deliver cleaner energy solutions, digital twins emerge as a powerful ally, offering a path to transforming traditional power production into a modern, interconnected, and environmentally conscious process.

Ultimately, digital twins represent a potent blend of innovation and practicality. They pave the way for the energy sector to not only meet current challenges but to anticipate and adapt to future demands. As the technology matures, its impact on optimizing power plants is likely to grow, driving the energy sector toward a brighter, more sustainable future. This transformation encourages technology enthusiasts, industry professionals, and decision-makers to reimagine what is possible, inspiring a new era of energy excellence.

Smart Grids and Renewable Energy

The energy landscape is rapidly evolving, driven by the imperative to transition towards sustainable sources and smarter consumption models. This evolution isn't just reshaping how we generate and consume energy, but it's also redefining the very infrastructure that underlies these processes. At the forefront of this transformation are smart grids—a sophisticated meld of digital technology and traditional energy systems, which stand poised to revolutionize the way we interact with electricity.

Smart grids operate as an intelligent network that enhances the existing electrical grid through the integration of digital communication technology. It's a shift from the unidirectional flow of electricity, where power simply traveled from production points to consumers, towards a multidirectional flow of information and energy. This dynamic system enables improved reliability, efficiency, and sustainability. By incorporating real-time data, smart grids can adjust and optimize electricity flow seamlessly across the network, addressing demand fluctuations, and minimizing energy waste.

There's a profound relationship between smart grids and renewable energy sources. Renewables, such as solar and wind power, are inherently variable and decentralized. They don't follow the predictable output patterns of traditional power plants. Thus, the flexibility of smart grids makes them ideal partners. For instance, digital twins of wind farms or solar panels can help predict power generation patterns based on weather forecasts, aligning energy dispatch with consumption needs in real-time. This harmony ensures a stable supply to the grid and mitigates the challenges associated with intermittency—a significant barrier to the widespread adoption of renewable energy.

Digital twins act as the digital counterparts of physical components within the grid, allowing for comprehensive simulation models and predictive analytics. These virtual replicas empower energy providers to test different scenarios without real-world repercussions, enhancing decision-making capabilities. Whether it's simulating the integration of a new renewable energy source or analyzing the impact of an outage, digital twins provide invaluable insights that preemptively solve problems.

The advantages of integrating digital twins into smart grids extend beyond logistical coordination. They offer a platform for predictive maintenance, reducing downtime and extending the lifespan of grid components. Instead of relying solely on scheduled upkeep, which can miss critical issues, digital twins help identify wear and tear through real-time monitoring. This proactive approach minimizes service disruptions and enhances the grid's reliability, which is especially crucial in areas dependent on renewables.

Beyond improving existing infrastructure, smart grids are also vital in empowering consumers. With smart meters, users gain insights into their energy consumption patterns. They're not just passive recipients of power; they become active participants. This shift towards prosumerism, where consumers also produce energy, is catalyzing the growth of microgrids. These are localized energy networks that can function independently or collaboratively with larger grids. They offer communities a sustainable and resilient energy option, often powered by local renewable sources, further decentralizing energy production.

The push for smart grids is also economic in nature. By optimizing energy flow and reducing losses, grids can significantly cut operational costs. Moreover, they open doors to new business models centered around energy as a service (EaaS). This model allows consumers to pay for energy services on a subscription basis, aligning with broader trends of servitization across industries. Digital twins again play a key role here; they help tailor these services to specific consumer needs, offering personalized energy solutions.

Another fascinating aspect of smart grids is their potential for facilitating a more sustainable future. As cities explore smart urban initiatives, the synergy between digital twins, smart grids, and renewable energy lays the foundation for smart cities. Urban environments can drastically reduce their carbon footprints by optimizing energy use across public transport, lighting, and even building energy management systems. The unification of these domains into a coherent system through digital twins ensures holistic sustainability efforts, paving the way for cities that are not only smart but also environmentally conscious.

Furthermore, in the context of climate change, the resilience of energy infrastructure becomes paramount. Smart grids, with their decentralized architecture bolstered by digital intelligence, are naturally more resilient to external disruptions, such as extreme weather events. By simulating potential disruptions and planning contingencies using digital twins, grid operators can better withstand adverse conditions and ensure swift recovery, safeguarding both electricity and economic stability.

Embracing this energy paradigm shift calls for robust policy and regulatory frameworks that support innovation while ensuring security and privacy. With more devices connected to the grid, cybersecurity becomes a critical concern. Ensuring data integrity and protecting infrastructure from potential cyber threats cannot be overstated. Hence, collaboration between regulatory bodies, energy providers, and technology developers is essential, setting standards that foster both innovation and protection.

In conclusion, smart grids represent a transformative approach towards an energy-efficient future. They bridge the gap between traditional energy systems and the increasing demand for sustainable, reliable, and flexible energy solutions. As digital twins continue to advance, the alignment between technology and energy systems grows even stronger, unlocking unprecedented opportunities for innovation. The foundation these technologies lay won't just support our current energy needs but also anticipate future demands, ensuring a sustainable tomorrow.

Chapter 8: Construction and Real Estate

The construction and real estate sectors are harnessing the power of digital twins to reshape how projects are envisioned and executed from inception to completion. These virtual models serve as dynamic companions to physical assets, offering a real-time, holistic view of project progress and performance. This technology enhances project management by enabling teams to simulate construction processes, predict future outcomes, and mitigate risks before they escalate into costly setbacks. Moreover, digital twins are becoming instrumental in boosting building performance through the lifecycle of the asset, fostering sustainable practices and operational efficiency. By integrating data from IoT devices and advanced analytics, stakeholders can optimize maintenance, energy consumption, and occupant comfort. Embracing digital twins in construction and real estate isn't just a trend but a strategic move toward a smarter, more resilient built environment that aligns innovation with sustainability imperatives.

Efficient Project Management

In the construction and real estate sectors, effective project management is essential to success. Projects are massive, complex undertakings that involve numerous stakeholders, tight deadlines, and substantial financial commitments. Digital twins have emerged as a transformative tool, enabling stakeholders to manage projects with increased efficiency and precision.

At its core, a digital twin is a highly detailed virtual model of a physical counterpart. In construction, this could mean a model of a building, bridge, or entire urban landscape. These digital replicas allow project managers to simulate and analyze different scenarios, facilitating informed decision-making that can prevent costly mistakes. By visualizing every aspect of a project, from materials used to the order of construction phases, stakeholders can achieve a clarity that static models simply can't provide.

Implementing digital twins in construction projects starts at the design phase. Architects and engineers use them to create 3D models that are more than just visual representations; they are dynamic entities that can be manipulated and tested under various conditions. For example, engineers can simulate the building's response to environmental stressors like wind or earthquakes, ensuring structural integrity before breaking ground. These intelligent models become a single source of truth, from which every team member operates, reducing discrepancies and misunderstandings.

One of the most critical aspects of project management is scheduling, and this is where digital twins offer substantial benefits. With all project data centralized in one place, including logistics, resource allocation, and timelines, digital twins facilitate more accurate project scheduling. Project managers can run simulations to predict potential bottlenecks, evaluate alternative strategies, and allocate resources more effectively. This predictive capability transforms project management from a reactive to a proactive discipline, allowing stakeholders to anticipate challenges and develop strategies in advance.

The integration of IoT devices with digital twins further enhances project management capabilities. Sensors placed on-site feed real-time data back into the digital twin, allowing for the continuous updating of the model with current conditions. This constant flow of information enables project managers to monitor progress, detect deviations from the plan, and act swiftly to correct course. Imagine knowing precisely when a delay in material delivery could impact the timeline or identifying unexpected issues in construction before they escalate. The digital twin becomes a central hub for continuous improvement and risk management throughout the project lifecycle.

Efficient communication and collaboration are fundamental to successful project management. Digital twins serve as a collaborative platform, bridging the gap between diverse teams and stakeholders. Whether on-site workers, remote engineers, or stakeholders across the globe, everyone gains access to the same up-to-date model, fostering understanding and enhancing teamwork. Decisions no longer rely solely on lengthy meetings or memos but stem from interactive, data-driven discussions facilitated by the digital twin.

Moreover, with augmented reality (AR) and virtual reality (VR) integrations, digital twins offer ground-breaking visualization and training capabilities. Project managers and team members can engage with the digital twin in immersive environments, walking through virtual job sites, assessing safety protocols, or planning complex installation tasks. This immersive interaction bolsters training, ensuring that teams are better prepared and can execute projects with higher proficiency.

Managing project costs is another area where digital twins make a significant impact. By providing more accurate estimates and allowing for detailed financial projections, digital twins help project managers adhere to budgets. Cost overruns, often caused by unforeseen changes or inadequate initial assessments, can be minimized. Digital twins offer the ability to simulate the financial implications of various scenarios, promoting a thorough analysis of the cost-benefit ratio of design choices or changes in project scope.

Digital twins also enable the adoption of predictive maintenance strategies during the construction phase. By evaluating equipment performance data through the digital twin, managers can predict when machinery might fail or require service, reducing downtime and associated costs. This foresight applies not only during construction but extends into the building's operational phase, providing a seamless transition from project management to facility management.

Beyond facilitating current projects, digital twins provide valuable insights that can be harnessed for future endeavors. As data accumulates, patterns and lessons emerge, enabling continuous improvement in construction practices. This historical data repository allows for benchmarking and the refinement of processes, ensuring that each subsequent project benefits from past learnings.

The potential of digital twins extends beyond technical functionalities; they play a crucial role in stakeholder engagement. As impressive communication tools, they can be used to showcase project progress to investors, government agencies, and the public. Enhanced visualization capabilities allow for transparent communication, aiding in gaining the confidence and support of all involved parties.

While integrating digital twins into project management brings numerous benefits, it does come with challenges. Strategic planning, investment in appropriate technology, and skilled personnel are essential to maximize their potential. Organizations must foster a culture that embraces digital transformation, encouraging upskilling and innovation among the workforce.

The future of efficient project management in construction and real estate lies in the capabilities and evolution of digital twins. By adopting this technology, projects can move closer to perfect synchronization, prediction, and collaboration. The result is more reliable, cost-effective, and timely project completions, a boon for developers, investors, and communities alike.

In conclusion, digital twins are not just a technological advancement; they are a paradigm shift in project management, opening doors to unprecedented efficiency and innovation in the construction and real estate sectors. As these industries continue to evolve, digital twins will undoubtedly remain at the forefront, driving progress into uncharted territories.

Enhancing Building Performance

Digital twin technology has emerged as a groundbreaking tool for enhancing building performance, offering capabilities that go far beyond traditional design and construction methods. By creating a virtual replica of a building or a system within it, stakeholders can simulate, analyze, and optimize performance in ways that were previously unimaginable. This leap forward is driven by the integration of advanced data analytics, the Internet of Things (IoT), and artificial intelligence (AI) technologies, which together provide unparalleled insights into building operations.

One of the most significant advantages of digital twins in the construction and real estate sectors is the ability to monitor and adjust a building's energy efficiency in real time. Buildings are complex systems with numerous variables impacting energy consumption, from HVAC systems to lighting and elevator use. By continuously receiving data from these systems, a digital twin can identify patterns that lead to energy waste and suggest corrective measures. This proactive approach not only leads to cost savings but also supports sustainability goals by reducing the building's carbon footprint.

For example, during peak energy consumption periods, a digital twin can simulate various scenarios to determine the best strategies for load shifting or demand response. By integrating weather data predictions, occupancy trends, and even local energy tariffs, building managers can optimize energy use dynamically. Moreover, these adjustments are not just theoretical; they can be implemented in real time, allowing for immediate benefits.

In terms of maintenance and operational efficiency, digital twins offer a paradigm shift. Through continuous monitoring, potential issues within a building's infrastructure can be detected before they become critical. Predictive maintenance becomes a reality, wherein algorithms analyze data from building systems to predict failures and suggest timely interventions. This shift from reactive to proactive maintenance enhances building performance and extends the lifespan of critical systems.

Another exciting application of digital twins is in improving indoor environmental quality (IEQ), a crucial factor for occupant health and productivity. A building's digital twin can track parameters such as air quality, light levels, temperature, and humidity, providing insights into how these affect occupants. Automated adjustments to HVAC and lighting systems ensure optimal conditions, contributing not only to comfort but potentially enhancing occupant performance and well-being.

Digital twins also offer substantial benefits during renovation and retrofitting projects. Virtual models allow architects and engineers to simulate structural changes, optimizing designs before any physical work begins. By understanding how changes might impact the building's systems and occupants, stakeholders can make informed decisions that save time and resources. This capability is particularly valuable in complex projects where multiple systems interact and change repercussions are difficult to predict.

The integration of digital twins with IoT devices creates an ecosystem where each component of a building is interlinked, sharing data that informs the virtual model. This real-time flow of information allows for insights that empower building managers to optimize space utilization. For instance, meeting rooms and communal areas can be monitored for occupancy patterns, enabling efficient use of spaces that might otherwise lay idle, ultimately enhancing the value of a property.

Digital twins also play a pivotal role in ensuring compliance with building codes and regulations. By simulating different scenarios, stakeholders can quickly ascertain if a building's current or proposed state meets necessary regulatory standards. This capability reduces the risk of costly rework and penalties, streamlining the approval process for modifications and new constructions.

Future advancements in digital twin technologies promise even more sophisticated building management. As machine learning algorithms improve, the predictive power of digital twins will grow, offering deeper insights into complex building systems and user interactions. This evolution will push the limits of what building performance enhancement can achieve, paving the way for smarter, more adaptive environments.

The journey toward adopting digital twin technology in construction and real estate is not without its challenges. Initial costs and the complexity of integrating new systems may deter some stakeholders. However, the long-term benefits—ranging from energy savings to extended asset life and enhanced occupant satisfaction—make a compelling case for investment. As more success stories emerge, hesitation is likely to give way to enthusiasm, heralding a new era in building performance optimization.

The digitization of the construction and real estate industries through digital twins represents a monumental step forward, akin to the industrial revolution of past centuries. As technology continues to evolve, the potential for enhanced building performance only grows, offering not just efficiency gains but a radical transformation of how living and working spaces are conceived and managed. The first movers in adopting these technologies will likely set new standards for innovation and excellence, inspiring a broader industry shift.

Chapter 9: Aerospace and Defense Applications

The aerospace and defense sectors are soaring to new heights with the transformative power of digital twins, radically reshaping the way aircraft and defense systems are designed and managed. By fostering innovative approaches to aircraft design and testing, digital twins allow engineers to create virtual models that accurately simulate physical systems. This leads to accelerated prototyping, safer test environments, and ultimately, more efficient production processes. In addition, digital twins enhance maintenance and lifecycle management by providing comprehensive data insights that predict potential failures before they occur, significantly reducing downtime and costs. The result is a formidable fusion of precision and adaptability, enabling the aerospace and defense industries to push boundaries while safeguarding investments and ensuring performance integrity. As digital twins continue to evolve, they promise to be the linchpins of a new era of aerospace and defense innovation, setting the stage for unprecedented technological advancement.

Aircraft Design and Testing

The integration of digital twins in aircraft design and testing is not just a technological advancement; it's a breakthrough that revolutionizes the aerospace domain. By simulating real-world scenarios in a digital environment, designers and engineers gain unprecedented insights into how aircraft will perform under diverse conditions. It's akin to having a crystal ball that predicts future outcomes, allowing for refined design processes and meticulous testing stages. This approach leads to not only enhanced performance but also greater safety and efficiency.

Digital twins provide a dynamic platform where every aspect of an aircraft's operation can be tested and optimized. From aerodynamics to structural integrity, the virtual model can mimic and analyze complex interactions. For example, turbulence impact on a new wing design can be precisely evaluated without the need for physical prototypes, greatly reducing both cost and time investments. The ability to iterate on designs rapidly accelerates innovation, pushing the boundaries of what's possible in aircraft engineering.

One of the groundbreaking moves in aircraft design has been the ability to perform virtual wind tunnel tests. These tests are vital for understanding airflow over the aircraft's surfaces and refining their aerodynamic efficiency. Digital twins take this a step further by allowing continuous modifications and real-time feedback on simulations, offering insights into potential improvements that would be inconceivable through traditional methods. This digital approach ensures a higher fidelity in results, with adjustments made in record time to achieve optimal performance.

In testing phases, digital twins have transformed how safety is approached. Advanced simulations can replicate extreme conditions—be it sudden weather changes, additional weight loads, or unexpected mechanical failures. By examining these scenarios virtually, engineers can develop robust aircraft systems that anticipate and compensate for real-world variables. This predictive capability significantly reduces risks associated with human trials and ultimately leads to the safer deployment of new aircraft.

Moreover, digital twins enable a holistic approach to system integration. Modern aircraft are complex machines with interconnected subsystems, such as avionics, propulsion, and communication. Testing these systems together in a virtual environment ensures that they function harmoniously, minimizing the risk of system failures when they finally come together in reality. This integrated testing ensures coherence and communication among systems are at their peak, ultimately delivering a seamless and resilient aircraft performance.

As interconnectedness becomes the norm, the integration of AI and IoT into digital twin technology further enhances aircraft design and testing. AI algorithms can sift through enormous datasets generated during simulations, identifying patterns and proposing optimizations that humans might overlook. IoT devices keep the digital twin updated with real-time data, enhancing the accuracy and reliability of simulations. This symphony of technologies fosters a new era of intelligent aircraft that learns and adapts over its lifecycle.

Beyond the technical aspects, the adoption of digital twins in aircraft design facilitates a shift in strategy and philosophy. It encourages cross-disciplinary collaboration, bringing together engineers, designers, data scientists, and software developers. This synergy of diverse expertise fuels innovative approaches and empowers teams to tackle challenges from multiple angles. It's not just about building a better plane; it's about fostering a culture that thrives on collaboration and innovation.

Additionally, digital twins offer a significant impact on sustainability in aircraft design. The simulations enable exploration of fuel-efficient designs or novel propulsion systems with minimal environmental impact. Testing these eco-friendly solutions digitally reduces the carbon footprint associated with traditional testing methods and aligns the aerospace industry with global sustainability goals. This approach not only meets regulatory demands but also satisfies consumer expectations for greener technology.

Digital twins also pave the way for customization in aerospace design. Aircraft can be tailored to specific customer needs, with digital models predicting the performance of unique configurations or materials. This flexibility allows airlines to optimize fleet capabilities, enhance passenger experience, and align with brand identity, ultimately gaining a competitive edge in the market. Testing these customizations virtually ensures they meet the highest standards before any physical changes occur.

The evolution of aircraft design and testing through digital twins is a testament to the transformative power of technology. As we continue to explore this digital horizon, the potential applications and benefits will only grow. The fusion of creativity, technology, and engineering not only demystifies digital twins but inspires their broader adoption throughout the aerospace sector. This trajectory promises to usher in a new age of aeronautics where innovation knows no bounds and where the sky is truly not the limit.

Maintenance and Lifecycle Management

In the rapidly evolving landscape of aerospace and defense, the application of digital twins for maintenance and lifecycle management emerges as a game-changing innovation. Let's explore how digital twins are transforming the way we think about aircraft maintenance and lifecycle optimization. No longer are these processes bogged down by lengthy inspections or reactive measures. Instead, digital twins offer a predictive, proactive approach grounded in real-time data and advanced simulations.

Digital twin technology facilitates comprehensive monitoring and diagnostics by creating a virtual replica of the aircraft. This replica integrates data from the aircraft's entire lifecycle, including design, manufacturing, and daily operations. This holistic view is crucial for pinpointing potential issues before they escalate, thereby ensuring operational readiness and maximizing asset uptime. More than just visualization tools, digital twins bring data to life, supporting informed decision-making and extending the service life of aerospace assets.

Imagine an air force base where every aircraft is paired with its digital twin. This digital counterpart continuously collects and analyzes data such as engine performance, structural integrity, and environmental exposure. It reflects the current state of the physical entity and predicts future conditions. Maintenance crews can access these insights to perform condition-based maintenance, sharply reducing the need for scheduled checks and allowing for targeted interventions.

The capability of digital twins in lifecycle management isn't limited to maintenance optimization. It provides invaluable feedback to the design and engineering phases. By continuously feeding real-world data back into the design loop, digital twins enable iterative improvements in aircraft models. Engineers can refine their designs based on actual usage patterns, material fatigue data, and unforeseen stressors that go unnoticed during initial testing phases. This creates a feedback loop that enhances product development with each iteration.

We've all witnessed the substantial costs associated with unscheduled maintenance and unexpected failures. Reducing these occurrences has always been a priority for the aerospace industry, but now digital twins offer a powerful solution. Imagine airplanes that communicate with their digital twins mid-flight, transmitting operational data that forecasts wear and tear. Ground crews can prepare for immediate maintenance as soon as an aircraft lands, turning downtime into a well-coordinated maintenance operation.

This approach not only cuts costs but also enhances safety. By addressing vulnerabilities before they manifest, digital twins help avoid catastrophic failures. When used in a defense context, where asset downtime can have critical implications, the value of such predictive maintenance becomes even more pronounced. Commanders can be confident in their fleet's readiness, knowing each piece of equipment is continually monitored and optimized.

Furthermore, the integration of AI and machine learning with digital twins accentuates their utility. These technologies help in analyzing vast datasets to uncover patterns and anomalies unseen by the naked eye. AI algorithms can adapt over time, improving their predictive accuracy with every dataset they process. As a result, the digital twin becomes an ever-evolving tool that reflects the latest in technological advancements and methodological improvements.

However, implementing digital twins in maintenance and lifecycle management isn't without its challenges. The complexity of integrating systems across different platforms, the management of vast amounts of data, and ensuring cybersecurity are significant concerns. But the potential benefits far outweigh these hurdles, driving innovation and investment in this arena.

Industry leaders must also address the human element. Maintenance crews and engineers need to be trained not only to use digital twins but to trust the insights they provide. This involves a cultural shift, recognizing that digital tools can complement and enhance human expertise. It also means developing new skill sets and fostering interdisciplinary collaboration to fully harness the potential of digital models.

In the end, digital twins in aerospace maintenance and lifecycle management embody a blend of cutting-edge technology and strategic foresight. They herald a future where every asset is perpetually optimized, downtime is minimized, and the life of expensive equipment is significantly extended. As enthusiasts, professionals, and decision-makers in this field, recognizing and championing these advancements will lead the way to unparalleled growth and innovation. Indeed, digital twins are not merely a tool but a transformative force reshaping the aerospace landscape, ushering in a new era of efficiency and reliability.

Chapter 10: Retail and Consumer Insights

In the ever-evolving landscape of retail, digital twins have emerged as a transformative force. As retailers seek more personalized shopping experiences, these virtual replicas harness data to unlock deep consumer insights, making interactions more engaging and relevant. Picture a seamless blend of online and offline experiences, where digital twins anticipate consumer desires and optimize every stage of the shopping journey. It's about transcending traditional retail challenges, and with inventory and supply chain optimization, the promise of minimized stockouts and overstock becomes a reality. Embracing this technology offers retailers a dynamic toolkit, not just for today's competitive market but also for crafting the future of consumer engagement. As they integrate these insights into their strategies, retailers drive innovation that not only caters to current trends but also sets the stage for the evolution of shopping itself.

Personalizing the Shopping Experience

In today's fast-paced digital world, retail is undergoing a renaissance, driven by groundbreaking technologies that are reshaping how consumers interact with brands. Among these innovations, digital twins stand out as a transformative force in personalizing the shopping experience. From virtual store replicas to personalized product recommendations, digital twins offer a deeper understanding of customer behavior, providing retailers with cutting-edge tools to enhance customer satisfaction and loyalty.

The concept of personalizing the shopping experience is not new. For decades, businesses have strived to tailor their offerings to meet individual customer needs. However, the advent of digital twins has taken personalization to a whole new level. By creating virtual counterparts of physical entities, retailers can now gather comprehensive data and insights into consumer preferences, often in real time. This enables them to dynamically adapt product offerings, store layouts, and marketing strategies to align with customer desires.

Imagine stepping into a store where every product on display is precisely what you've envisioned in your mind. With digital twins, this vision isn't as far-fetched as it once seemed. Through real-time data collection from various sources—such as mobile apps, in-store sensors, and online behaviors—retailers are developing a comprehensive view of each customer's journey. This holistic perspective allows them to anticipate shopper needs and offer personalized solutions before a purchase decision is even made.

Furthermore, digital twins empower retailers to optimize store operations, create immersive customer experiences, and tailor marketing campaigns. By maintaining a digital replica of a physical store, businesses can simulate any number of scenarios—from changing product placements to adjusting inventory levels based on predicted demand. These simulations inform strategic decisions that ensure customers find what they need, when they need it, with minimal effort.

Personalizing customer interaction doesn't stop at planning and logistics. The augmentation of shopping environments with digital twins extends into customer engagement, offering features like interactive displays and virtual fitting rooms. These innovations allow shoppers to visualize products in ways that were previously unimaginable, whether trying on clothes virtually or previewing furniture in their own homes through augmented reality. Retailers can effectively double down on customer engagement by analyzing which digital experiences resonate most with their consumers.

The symbiotic relationship between digital twins and artificial intelligence (AI) algorithms is particularly profound in the retail sector. AI plays a crucial role in analyzing the data curated by digital twins, shedding light on complex consumer behavior patterns and extracting actionable insights. This, in turn, enables predictive personalization—where algorithms can recommend products or services tailored precisely to individual tastes and preferences.

Shopping today extends well beyond transactional events; it's about crafting memorable experiences. By leveraging the power of digital twins, retailers can imagine environments where every shopping experience feels as though it was designed exclusively for the individual. As digital twins map out detailed consumer profiles, retailers can deploy highly targeted marketing campaigns, enhancing customer engagement and retention.

While the benefits are plentiful, the implementation of digital twins in retail is not without challenges. Data privacy concerns are at the forefront as retailers collect and process large amounts of consumer data to feed their digital twin systems. To address these issues, businesses must prioritize ethical data management practices, ensuring transparency and trustworthiness in their engagements with customers.

Despite these challenges, the potential of digital twins in elevating retail experiences is undeniable. They offer a blueprint for the future, where personalized shopping transcends mere recommendations and encompasses fully tailored interactions from start to finish. As the retail landscape continues to evolve, those who embrace digital twin technology will likely lead the charge, shaping new standards for personalized shopping in the digital age.

In conclusion, the integration of digital twins into the retail sector marks a paradigm shift towards more personalized and engaging shopping experiences. By embracing this technology, retailers unlock the potential to not only meet the evolving demands of today's consumers but to exceed their expectations. As digital twin technology advances, the line between the physical and virtual realms continues to blur, transforming the very fabric of how we shop and interact with brands.

Inventory and Supply Chain Optimization

Retail has always been a dynamic industry, constantly adapting to consumer demands and technological advancements. The introduction of digital twins opens new vistas for inventory and supply chain optimization, tapping into real-time data and predictive analytics to streamline operations. Simply put, digital twins are virtual replicas of physical entities, systems, or processes, enabling a two-way interaction between the real world and its digital counterpart. In the context of retail, these twins offer unparalleled visibility into inventory and supply chain functions, facilitating timely decision-making and enhanced operational efficiency.

The race to keep shelves stocked without overloading warehouses is a perpetual challenge for retailers. However, digital twins can provide a single source of truth by creating transparent inventories and supply chain data flows. They can leverage data from a myriad of sources—such as RFID tags, IoT sensors, and point-of-sale systems—to construct a comprehensive view of stock levels and movement patterns. This not only helps retailers match supply with demand more accurately but also reduces the risks associated with overstocking or stockouts, which are critical for maintaining profit margins in a competitive market.

Predictive analytics is another frontier where digital twins exhibit their strength. By simulating different scenarios, companies can anticipate potential disruptions in the supply chain before they manifest in reality. Whether a sudden surge in demand or an unforeseen delay in shipment, digital twins allow companies to model the impact of these variables and create strategic plans to mitigate their effects. Such foresight helps retailers preempt issues, ensuring that customer satisfaction remains high by delivering products on time and in adequate quantities.

Logistics and transportation are critical aspects of the supply chain that stand to gain significantly from digital twin integration. The process encompasses the journey of goods from production facilities to end consumers, involving numerous intermediate steps such as warehousing, distribution centers, and retail outlets. Digital twins facilitate route optimization, load balancing, and real-time tracking of goods in transit. As a result, they contribute to reduced transportation costs, decreased delivery times, and minimized carbon footprints—crucial factors in an increasingly eco-conscious market.

But where digital twins could truly reshape retail is in optimizing the intricate web of supplier networks. Consider the complexity involved in managing thousands of suppliers globally, each with its own unique lead times, capacities, and constraints. Digital twins can dynamically adjust procurement strategies based on up-to-the-minute data, aligning supply chain operations closely with market demands. They enable a more agile approach, replacing traditional, rigid supply chains with flexible, efficient networks capable of responding swiftly to fluctuations in the market landscape.

Risk management is another critical benefit offered by digital twins in supply chain optimization. In the face of natural calamities, geopolitical tensions, or other unforeseen disruptions, digital twins can simulate various risk scenarios to enhance resilience and continuity. By identifying vulnerabilities within the supply chain, businesses can devise contingency plans that ensure minimal disruption to operations, guarding against financial losses and maintaining brand reputation.

Moreover, digital twins are pivotal in sustainability efforts. As consumers become more environmentally conscious, retailers are under pressure to adopt greener practices. Digital twins enable precise inventory management by analyzing product lifecycle data, thereby minimizing waste through better forecasting and planning. Coupled with supplier network optimization, they support ethical sourcing and reduce the carbon footprint associated with logistics, aligning with global sustainability goals.

Leveraging digital twins comprehensively requires collaboration between IT and operational teams, necessitating a shift in organizational culture. Such collaboration underlines the importance of data integration and secure sharing practices across departments. It's crucial for retailers to invest in training programs that equip employees with the skills necessary to utilize digital twin technology effectively, fostering an environment where teams can innovate and drive business value.

The pathway to adopting digital twins in retail must acknowledge existing technological infrastructure, as well as potential roadblocks in data integration and legacy systems. Transitioning to a twin-enabled model demands an iterative approach, involving pilot projects and phased deployments to build confidence and drive value gradually. As the technology matures and proves its worth, scaling operations while mitigating initial implementation challenges becomes progressively easier.

Ultimately, the implementation of digital twins in inventory and supply chain optimization serves not just as a technological upgrade but as a fundamental transformation. It infuses agility, efficiency, and transparency into the retail sector, empowering firms to meet consumer expectations with precision and reliability. As digital twins continue to advance, they hold the promise of redefining how retailers perceive and manage their supply chains, becoming an essential component of the modern retail ecosystem.

Chapter 11: Agriculture and Food Production

In recent years, the agriculture sector has embraced the transformative power of digital twins to revolutionize traditional farming practices. We're witnessing a shift from conventional methods towards precision farming, where virtual models simulate and analyze real-time data. These digital replicas of fields and crops allow farmers to make informed decisions, optimize yield, and reduce waste. The integration of sensors and IoT devices enables real-time monitoring of soil health, weather patterns, and crop conditions, fostering a proactive approach to cultivation. This technological leap not only enhances productivity but also aligns with sustainability goals, conserving resources like water and energy. As we innovate, digital twins present an inspiring vision for food production systems that are more efficient, resilient, and environmentally sound, paving the way for a future where technology and agriculture coexist harmoniously to feed the world's growing population.

Precision Farming Techniques

Precision farming, often dubbed precision agriculture, is revolutionizing how we view the fields and farms of today. This approach uses cutting-edge technology to monitor conditions within fields, analyzing everything from soil composition to real-time weather patterns. The goal is clear: maximize crop yield while minimizing waste, ultimately creating a sustainable agricultural practice. Precision farming empowers farmers with actionable insights, enhancing decision-making processes and transforming traditional practices into data-driven strategies.

One of the foundational aspects of precision farming is its reliance on satellite imagery and GPS technology. By employing these tools, farmers can create detailed maps of their fields, highlighting variations in crop health, moisture levels, and soil quality. This capability doesn't just allow identification of problematic areas; it also informs targeted interventions. Rather than applying fertilizers and pesticides uniformly and often wastefully, farmers can apply the right amount precisely where it's needed. This floor of focused applications conserves resources and protects the environment, reducing runoff into nearby water bodies.

Machine learning and artificial intelligence (AI) are playing pivotal roles in advancing precision farming techniques. These technologies sift through the vast amounts of data collected from fields, analyzing patterns and predicting outcomes. For instance, AI can help predict weather conditions and their potential impact on crops, enabling farmers to plan their activities accordingly. By leveraging historical weather data and combining it with real-time updates, AI provides agricultural professionals with unmatched foresight.

Sentinel systems, which comprise sensor networks distributed across fields, offer staggering amounts of real-time data. Sensors monitor variables like soil moisture, soil temperature, and nutrient levels. This real-time feedback loop allows farmers to adapt to changes swiftly. Computational models utilize this data to simulate numerous scenarios, predicting crop yields and water usage accurately. Through these models, farming becomes an adaptable dance with nature, allowing for timely and informed decisions.

Variable rate technology (VRT) is an extension of precision farming that acts as a game-changer for resource management. VRT systems adjust the application rates of inputs like seeds, fertilizers, and irrigation in response to the specific requirements of each field section. Such meticulous application leads to a striking decrease in wasted resources while simultaneously improving the quality and quantity of the yield. The ability to tailor farming practices at such a granular level not only enhances efficiency but also supports sustainable farming by reducing environmental impact.

Precision farming isn't limited to just data collection and analysis. The implementation of robotics is becoming more prevalent within the sector. Autonomous tractors and drones represent the next leap, promising substantial changes in how agriculture is managed. These machines can perform tasks such as planting, weeding, monitoring, and harvesting. In regions where labor shortages are common, agricultural robots offer much-needed relief by taking over repetitive and labor-intensive tasks. Drones equipped with multispectral imaging cameras can scan fields, providing visual data that's processed to reveal insights unseen to the naked eye.

While technology plays a significant role, the importance of collaboration in precision farming can't be overstated. Farmers, agronomists, data scientists, and policymakers must work together to streamline the integration of precision technologies into day-to-day operations. Partnerships with tech companies, research institutions, and governmental bodies can accelerate innovation and make these technologies more accessible to small and medium-sized farms. By establishing a cohesive network of knowledge and support, precision farming can be cultivated to benefit larger communities and economies.

In this rapidly evolving landscape, continuous training and education are paramount. Bringing technology literacy to farming communities ensures they can fully utilize modern tools and techniques. Workshops, training sessions, and online courses bridge the gap between traditional farming methods and the digital advancements that define precision agriculture today. Fostering an environment of learning and adaptation prepares farmers to not only meet current challenges but to embrace future innovations with confidence.

Despite the many advantages, precision farming techniques do face certain challenges. High initial costs for equipment and technology can be prohibitive, especially for smaller farms. However, as technology evolves and becomes more commonplace, costs are gradually decreasing, making precision farming more accessible. Governments and financial institutions can also support this transition by providing subsidies and favorable loan terms to encourage adoption.

Moreover, the ethical considerations involved in data ownership and privacy are complex. As farms become data-rich environments, questions about who owns the insights and how they're used become critical. Establishing clear guidelines and protections for data will be essential as precision farming continues to expand. Farmers must feel secure in sharing their data, knowing it will be used responsibly and to their benefit.

Precision farming represents a paradigm shift, yet it maintains the heart of agriculture—the deep-rooted connection with the land and its bounty. It is a testament to how technology, when wielded with care and insight, can honor tradition while paving new pathways towards efficiency and sustainability. By nurturing technology and soil in equal measure, we step into a new era of agricultural achievement—one that holds promise not just for better yields, but for a better world.

Sustainability and Resource Management

As we're facing increasing global challenges in terms of food security, climate change, and resource limitation, the agricultural sector stands at a pivotal point in its evolution. Integrating digital twins into agriculture and food production offers a promising pathway toward sustainability and efficient resource management. This might sound bold, but it's true; digital technology could unlock unprecedented potential for the agricultural industry by refining processes, reducing waste, and enhancing productivity.

Imagine a world where farmers have the ability to virtually replicate entire ecosystems, allowing them to test sustainability strategies in a risk-free environment. Digital twins make this possible. By creating exact virtual copies of fields and resources, they enable the simulation of different scenarios. Farmers can experiment with diverse combinations of crops, irrigation levels, and soil treatments—all without disrupting the physical fields. This helps in making data-driven decisions that not only improve yield but also contribute to environmental conservation.

Resource management, particularly water and soil health, is crucial for sustainable agriculture. Water scarcity is an ongoing concern, and over-irrigation can lead to soil degradation. Here, digital twins make their mark by providing precise measurements and predictive analytics. These models analyze the required water content for specific soil and plant types. Consequently, they optimize irrigation schedules to minimize water waste. Real-time data is essential; with sensors providing constant feedback, digital twins adjust strategies dynamically to evolving conditions, ensuring resource conservation without compromising crop health.

Moreover, digital twins in agriculture play a key role in enhancing renewable energy integration. Farms can leverage digital models to design and optimize energy systems, such as solar panels or wind turbines. These systems can be tailored to meet the specific energy needs of agricultural operations. Digital twins analyze energy consumption patterns and environmental conditions, enabling farmers to deploy the most efficient mix of renewable technologies. Ultimately, this reduces the dependence on fossil fuels, cutting operational costs and carbon footprints.

Sustainability isn't just about conserving resources; it's about creating regenerative systems. Digital twins encourage regenerative agricultural practices by helping manage crop cycles and rotational grazing optimally. They help predict the best crop rotation schemes, avoiding monoculture pitfalls by ensuring soil nutrients are restored naturally. With detailed insights into crop performance and soil health, farmers can promote biodiversity and maintain ecosystem balance, thus supporting long-term soil fertility and resilience.

Let's not underestimate the power of data-sharing platforms powered by digital twins. These platforms foster collaboration among farmers, researchers, and policymakers. By sharing insights via digital twins, stakeholders can identify patterns, predict disruptions, and devise collective strategies for managing resources more effectively. It becomes a communal effort to tackle issues like pest infestations, climate unpredictability, and resource scarcity. This cooperative approach enhances resilience across the agricultural community, ensuring food security and sustainability at a larger scale.

Furthermore, accelerating the adoption of such advanced technologies might initially seem daunting for smallholder farmers due to perceived complexity and costs. However, the long-term savings from efficient resource management and increased productivity often outweigh initial investments. Training and awareness programs in cooperation with technology providers can ensure that digital twins are accessible and beneficial to all, promoting equitable development and sustainable growth.

Incorporating circular economy principles, digital twins can also influence how agricultural waste is handled. By simulating potential outcomes, farmers can devise innovative solutions to reuse waste, such as converting organic matter into bioenergy or fertilizer. These practices not only reduce waste but also create additional revenue streams and promote sustainability.

Ultimately, digital twins in sustainability and resource management signify a shift in how we perceive and interact with our agricultural systems. They offer a lens to view agriculture holistically, where science and technology empower farmers to become custodians of their land. By providing insights that drive precision and foresight, digital twins create a landscape where productivity and sustainability are harmoniously aligned.

As we continue exploring digital twins' capabilities in agriculture, we take significant steps towards a future where resource efficiency and sustainability are standard practices, securing both the planet's health and humanity's sustenance. This isn't just a technologic transformation; it's a call to imagine and construct an agricultural system that respects boundaries and thrives within them.

Chapter 12: Digital Twins in Telecommunications

As telecommunications rapidly evolves, the promise of digital twins emerges as a catalyst for innovation, driving the industry toward unparalleled efficiencies. By mirroring network components, digital twins offer a dynamic platform for simulating network behavior and understanding complex system interactions in real-time. This enables telecom providers to optimize network infrastructure planning and maintenance, ensuring seamless service delivery. Moreover, digital twins empower service providers to enhance customer experience by preemptively addressing issues, thus minimizing downtime. Imagine the ability to run comprehensive scenarios for network upgrades without disrupting existing services—this isn't science fiction but a practical application facilitated by digital twins. With these digital replicas, the telecommunications industry can boldly step into a future where agility and predictive capabilities redefine network management strategies, ultimately leading to smarter, more resilient communication networks.

Network Infrastructure Planning

In the rapidly evolving landscape of telecommunications, the concept of digital twins has emerged as a game-changing tool for network infrastructure planning. It provides a dynamic, virtual replica of physical network components and environments, allowing operators to visualize, simulate, and predict network behavior with unprecedented precision. Initially, the prospect of planning and adapting vast network infrastructures seemed daunting, but digital twins are transforming these challenges into manageable opportunities.

Telecommunications networks are the backbone of digital communication, and their complexity has grown significantly with the proliferation of devices and the exponential increase in data traffic. To manage this growth efficiently, network infrastructure planning has had to undergo radical transformation. The adoption of digital twins offers a visionary approach, enabling telecom operators to create a comprehensive digital representation of their entire network architecture, including switches, routers, data centers, and even the physical topologies of cables.

Imagine visualizing what-if scenarios in real-time, where telecom engineers can stress-test their networks under various simulated conditions without any risk to the actual systems. Digital twins possess the capability to simulate traffic variations, system failures, and new technology integrations. This simulation capability empowers network engineers to assess the impact of potential infrastructure expansions or optimizations before deploying them in the real world. The results are not just theoretical. By using a digital twin to analyze how a network would handle surges like those during a major sporting event or a cyber-attack, companies can plan proactive interventions that enhance reliability and performance.

Beyond immediate operational benefits, the strategic advantages of leveraging digital twins in telecommunications are profound. By harnessing historical and real-time data, these digital representations offer a far-reaching vision of infrastructure evolution over time. This foresight aids in long-term planning and investment strategies, ensuring that the network infrastructure can swiftly adapt to technological shifts such as 5G deployment and beyond.

Consequently, this proactive planning reduces costs associated with unplanned downtime, inefficient upgrades, and emergency repairs.

The notion of predictive maintenance, traditionally a staple in industries like manufacturing and automotive, has begun to resonate strongly within telecommunications. Digital twins enable predictive insights into when network components might fail, identifying patterns and anomalies that would be imperceptible to human inspectors. By anticipating issues before they manifest, service providers can allocate resources more effectively, prioritizing upgrades and maintenance in a way that minimizes disruption and maximizes efficiency.

Furthermore, digital twins challenge conventional approaches to network design and expansion. Traditional methods heavily rely on historical data and linear projections to estimate future demand. However, digital twins transcend these limitations, providing a multifaceted and dynamic view that encompasses real-world variables such as geographic, demographic, and even environmental factors. This capability ensures that network infrastructure not only meets current demands but is also resilient enough to handle unforeseen spikes in usage.

As digital twins handle an abundance of complex data, their integration into existing networks can seem overwhelming. Yet, the transition is often less cumbersome than anticipated. By adopting iterative integration strategies, telecom operators can gradually build digital twin capabilities into their systems. This phased approach allows for manageable increments of change, ensuring that technological advancements go hand-in-hand with existing operational procedures.

Crucial to network infrastructure planning through digital twins is the synergy between various technologies. Coupled with advancements in AI and machine learning, digital twins provide insights that are not merely descriptive but also prescriptive. AI algorithms, for instance, can analyze the data generated by a digital twin to suggest optimal paths for network traffic, further refining the efficiency of the entire system. This level of sophistication in data analysis offers telecom operators a powerful tool to stay ahead of the curve.

The constant evolution of network technology requires a robust and flexible approach to infrastructure planning. Digital twins offer the decentralization and flexibility needed to adapt a network rapidly without extensive overhaul, a critical factor when facing technological disruption. As the industry gravitates towards even more distributed networks facilitated by edge computing, digital twins stand at the forefront of redefining telecom infrastructure, making them indispensable to modern network strategies.

While the promises of digital twins are vast, the successful deployment in telecommunications relies on overcoming a series of operational and technical hurdles. Integration challenges, for one, require meticulous coordination across departments and stakeholders. Yet, these are not insurmountable. By investing in cross-disciplinary training and aligning organizational goals with digital strategies, telecom operators can streamline this transition, ensuring that new technologies are seamlessly woven into the fabric of their operations.

The transformative nature of digital twins in network infrastructure planning is undeniable. As telecommunications networks continue to grow in complexity, the insights and efficiencies gained from digital twin technologies offer a strategic edge. These virtual models are not merely a novel addition but a necessity for operators aiming to ensure robust, adaptive, and future-proof networks. By fully embracing the capabilities of digital twins, telecommunications can achieve a level of sophistication and adaptability that was once thought unimaginable.

Enhancing Customer Experience

The telecommunications industry thrives on connectivity and communication, serving as the backbone of modern society's digital interactions. In this landscape, digital twins are emerging as powerful tools to transform how service providers interact with their customers. By creating virtual replicas of network infrastructure and customer usage patterns, telecommunications companies can offer unprecedented levels of personalization and service optimization, enhancing the overall customer experience.

Digital twins enable telecommunications providers to delve deeper into customer behavior, identifying patterns and preferences that were previously obscured in vast datasets. With this insight, companies can tailor services to align exactly with customer needs, presenting personalized options for data plans, bundled services, and content delivery. Imagine a system that anticipates a user's need for increased data during a vacation and offers a just-in-time package tailored to those needs—digital twins make this a reality.

An integral feature of digital twins is their ability to facilitate proactive network management. By simulating various scenarios, providers can predict and mitigate potential service disruptions or degradation, thereby ensuring a seamless communication experience for customers. This proactive approach not only enhances service reliability but also instills customer confidence and satisfaction, potentially reducing churn rates significantly.

Customer support also stands to benefit tremendously. Digital twins facilitate a shift from reactive to proactive customer service. By analyzing the digital twin of a customer's connectivity, support teams can anticipate and resolve issues before the customer is even aware of them. For instance, if a digital twin predicts a potential drop in service quality due to congested nodes, a fix can be implemented preemptively, ensuring that the customer experiences uninterrupted service.

Moreover, digital twins can play a pivotal role in enhancing customer experience through the delivery of immersive and interactive services. As augmented and virtual reality applications become more prevalent, telecommunications providers can leverage digital twins to optimize the delivery of these bandwidth-intensive services. By ensuring optimal network conditions and simulating user experiences, digital twins enable the delivery of seamless and engaging AR/VR experiences.

Another noteworthy aspect is the facilitation of informed decision-making by customers. Digital twins can provide users with a window into their own usage patterns in a comprehensible format, empowering them to make better decisions regarding their service plans. Perhaps a user who frequently hits their data cap might benefit from a personalized notification and a recommendation tailored to their specific usage profile, all thanks to the insights acquired through digital twins.

Through integrating digital twins, telecommunications providers can also redefine how they handle network maintenance and upgrades. Rather than just reacting to network failures, providers can use digital twins to simulate changes and foresee their impact. This simulation-driven approach allows for a strategic roll-out of network upgrades, secure in the knowledge that customer experience won't be disrupted by unforeseen issues. The result is a more resilient network that continues to deliver high-quality service even as infrastructure evolves.

These advancements in customer experience are further enhanced by the potential for data-driven customization. Digital twins offer granular insights into network usage, helping service providers devise highly personalized marketing strategies. For instance, showcasing targeted offers directly aligned with individual consumer behavior improves the efficacy of marketing efforts and enhances customer satisfaction—everyone wins.

It's not just about proactive management or tailored marketing; digital twins open the door to innovative customer engagement activities. Think about augmented reality platform offerings where users engage with virtual environments that are responsive to real-time conditions. Such experiences can deepen user engagement, transforming ordinary interactions into captivating experiences.

Despite these capabilities, the transformative power of digital twins must be balanced with respect to privacy and ethical considerations. Customers will appreciate the enhancements in service quality and personalization, but their trust hinges on the transparent and secure handling of their data. Telecommunications providers must ensure robust data governance protocols to safeguard customer information while harnessing the potential of digital twins.

Ultimately, digital twins in telecommunications represent an exciting frontier for enhancing customer experience. As this technology matures, the possibilities for creating more satisfying, personalized, and reliable services continue to grow. For industry professionals and decision-makers, the challenge lies in leveraging these capabilities while maintaining a keen focus on safeguarding customer trust and fostering long-term relationships. In doing so, digital twins can truly become a linchpin in the quest to offer unparalleled customer experiences in the telecommunications arena.

Chapter 13: Navigating Challenges and Risks

As digital twins continue to redefine industries, navigating the challenges and risks they present becomes crucial. The rapid pace of technological evolution brings about complex issues surrounding data privacy and the ethical management of vast datasets. Balancing innovation with security requires a proactive approach, where decision-makers must be vigilant guardians of sensitive information while fostering a culture of transparency. Tackling these challenges isn't merely a technical endeavor; it also involves addressing societal concerns and ethical dilemmas that arise with unprecedented data capabilities. Moreover, as digital twins embody the confluence of diverse technologies, organizations must prepare for potential integration hurdles and recognize the intricate dance between opportunity and risk. Embracing these challenges head-on will not only safeguard trust but also pave the way for sustainable innovation that honors both technological advancement and human values.

Data Privacy and Security Concerns

As digital twins continue to transform industries from aerospace to healthcare, concerns about data privacy and security are escalating. These twin avatars of physical entities are not just technological wonders; they are significant repositories of data, capturing detailed insights that fuel innovation and optimize operations. However, they also present vulnerabilities that, if not addressed diligently, could turn dreams of transformation into nightmares of data breaches and privacy violations.

Understanding the scale of data involved is crucial. Digital twins rely on a wealth of information to simulate and predict outcomes in real time. This data typically includes sensitive information—be it patient health records in healthcare applications or proprietary operational data in manufacturing. The challenge lies in ensuring that all this data is securely handled and protected against unauthorized access. A breach could have severe implications, including financial loss, reputational damage, and regulatory fines.

The intersection of IoT and AI with digital twins brings additional layers of complexity to the security landscape. IoT devices continuously collect data, sometimes from remote and unsecured locations, making them prime targets for cyberattacks. AI algorithms process this data to refine the digital twin models, and any tampering with the data inputs can lead to faulty predictions and unintended consequences. Therefore, securing not only the digital twin but its entire data ecosystem becomes crucial.

One significant area of concern is the transmission of data between devices, models, and other systems. This data needs to be encrypted, and robust authentication protocols should be in place to prevent unauthorized interception. However, encryption alone is not a panacea. It must be complemented by intrusion detection systems that can alert stakeholders to potential breaches immediately and take corrective action without delay.

Access control plays a pivotal role in maintaining data privacy. Organizations must implement stringent access management policies that define who can interact with different aspects of digital twins. Role-based access controls and multi-factor authentication can help ensure that only authorized personnel have access to sensitive data and functionalities. Regular audits and monitoring of access logs also help in identifying any unauthorized attempts to gain entry.

Furthermore, while digital twins promise better decision-making through data-driven insights, these systems often require sharing of information across different organizations and departments. This begs the critical question: how can information flow seamlessly while still respecting privacy constraints? Implementing data governance frameworks that outline data usage, sharing policies, and accountability measures is a step in the right direction. Such frameworks should also align with existing privacy laws and regulations.

Certain sectors, such as healthcare, face heightened scrutiny regarding data privacy given the sensitive nature of personal health information. Regulations such as HIPAA in the United States demand stringent safeguards to ensure the confidentiality, integrity, and availability of personal health data. Violations can lead to significant penalties, not to mention a loss of trust. Digital twin implementations in such sectors need to prioritize compliance from the ground up, building privacy and security considerations into the initial design of systems.

Data anonymization is an additional strategy that can be employed to protect individual privacy while still leveraging broad datasets for insights. By stripping data of personally identifiable information, organizations can mitigate the risk of identity compromise. However, this must be done judiciously to ensure the utility of the data isn't also lost in the process.

From a technological standpoint, blockchain technology presents interesting possibilities for enhancing the security of digital twin ecosystems. By providing an immutable ledger of transactions, blockchain can help ensure data integrity and traceability, making it harder for malicious actors to alter datasets unnoticed. This story of technology aiding technology could become a cornerstone of future digital twin security architectures.

The importance of a robust cybersecurity culture cannot be overstated. This involves not just technological solutions but also a focus on human factors. Employees at all levels must be educated about cybersecurity best practices. Regular training sessions can ensure they are aware of potential threats and are well-equipped to respond promptly.

In the pursuit of a secure digital twin environment, collaboration becomes key. Industry stakeholders, cybersecurity experts, and policymakers need to work hand in hand to develop and implement industry standards that keep pace with technological advancements. Collaborative efforts can foster innovation in security solutions and ensure comprehensive coverage of potential vulnerabilities.

In conclusion, while digital twins hold unparalleled potential for innovation across sectors, they bring with them significant data privacy and security challenges that need proactive and multi-faceted responses. By prioritizing these aspects, organizations can not only safeguard their digital twin initiatives but also pave the way for trust-driven technological progress. The future depends on how well we can manage these invisible yet critical aspects of digital twins, ensuring they remain secure custodians of our data-driven world.

Ethical Considerations

As digital twins blaze new trails in various industries, they simultaneously traverse complex ethical landscapes that are both vast and nuanced. The convergence of virtual representations with real-world entities raises questions that are not just philosophical but also profoundly practical. Ethics, in this context, is not merely about following a moral compass; it's about understanding the ramifications of each technological step and its ripple effects on society.

One immediate ethical concern is data privacy. Digital twins rely heavily on data acquisition to create accurate simulations of real-world systems. Whether it's a smart city using digital twins for urban planning or a healthcare provider employing them for patient care, the data involved is often sensitive. It's vital to ask how this data is collected, who has access to it, and how it is stored and utilized. Building trust is essential, as public apprehension about data misuse can be a significant barrier to digital twin adoption. Without stringent privacy measures, the potential for breaches or misuse could undermine all the benefits these technologies offer.

Moreover, ownership and consent present ethical challenges. When digital twins of people—say, a patient's vitals in a healthcare setting—are created, who owns that digital representation? Is it the entity that developed the model or the person whose data powers it? These questions are not just hypothetical. They strike at the heart of personal sovereignty in an increasingly digital world. Establishing clear guidelines and frameworks for ownership is crucial to prevent the exploitation of individuals' data.

Algorithmic bias and inequity are also significant ethical considerations. Digital twins often utilize AI and machine learning for data processing and predictive analytics. If these algorithms are biased due to flawed training data, they can perpetuate or even exacerbate existing societal biases. For instance, a digital twin system used in managing urban infrastructure could inadvertently marginalize communities if it's designed using biased data. Addressing bias requires ongoing diligence and rigorous testing to ensure inclusivity and fairness.

The deployment of digital twins raises ethical concerns about transparency and explainability. When decision-makers use digital twins to inform policy or operational decisions, the reasoning behind these decisions must be understandable to stakeholders. This becomes especially critical when decisions have significant societal impacts. Transparent methodologies and clear communication about how digital twins derive their conclusions are necessary to maintain accountability and public trust.

Environmental and sustainability issues are ethical angles often overlooked in the rush towards technological innovation. Digital twins, especially those linked with energy-intensive computation, can contribute to environmental degradation if not managed sustainably. As stewards of these technologies, it's ethically imperative to consider their carbon footprint and strive for solutions that align with global sustainability goals. Balancing technological advancement with environmental conservation should be a guiding principle in the development and deployment of digital twins.

As digital twins become more prevalent in decision-making processes, one can't ignore the ethical implications of tech-driven unemployment. Automation and optimization processes powered by digital twins can lead to job displacements. Companies must weigh the efficiency gains against the societal costs of layoffs and prepare strategies to mitigate these impacts. This could include retraining programs to help displaced workers transition to new roles within the evolving digital economy.

The ethical landscape surrounding digital twins is further complicated when applied in sectors like defense and law enforcement. While digital twins can enhance capabilities and effectiveness, they also come with heightened risks of misuse. There is a thin line between ensuring security and infringing on personal freedoms. Robust ethical frameworks are necessary to guide the responsible use of digital twins in contexts where human rights and civil liberties could be at stake.

Balancing innovation with ethical responsibilities is not a straightforward task, but it is necessary for sustainable progress. By embedding ethical considerations into the fabric of digital twin technologies, stakeholders can ensure that these innovations are not just groundbreaking but also principled and socially responsible. The intersection of ethics and technology is a dynamic space that requires constant vigilance, dialogue, and adaptation as both fields evolve.

Inclusion in decision-making processes is key to addressing these ethical challenges. Developers, users, and communities impacted by digital twin technologies should all have a seat at the table. By fostering inclusivity and equitable participation, ethical considerations can be better identified and addressed, paving the way for a more holistic approach to technological advancement.

At the heart of all these considerations is the notion of ethical stewardship. Decision-makers and technology leaders must look beyond short-term gains and consider the long-term implications of their actions. Building a culture of ethics in technology development not only ensures responsible innovation but also positions digital twins as tools for good, bridging the gap between the digital and physical worlds in ways that enrich rather than exploit.

Chapter 14: The Role of Cloud Computing

Cloud computing has become indispensable in the digital twin ecosystem, catalyzing seamless integration and unparalleled scalability. As our reliance on data continues its exponential rise, the cloud offers a sanctuary—transforming potential data overload into an opportunity for rich insights. By empowering organizations to store vast amounts of data efficiently, cloud platforms facilitate the real-time processing and analytics necessary for digital twins to reflect and respond to their physical counterparts dynamically. This elasticity, available at scale, democratizes access to digital twins, paving the way for innovation regardless of an organization's size. The cloud's role extends beyond mere storage; it's a backbone for the diverse applications and services that digital twins rely on, ensuring that decision-makers can act swiftly and with foresight. As we harness this robust computational resource, cloud computing ensures that the digital transformation journey with digital twins is both scalable and sustainable, driving future forward-thinking innovations.

Seamless Integration and Scalability

The emergence of cloud computing has been a transformative force in numerous industries, but it's particularly significant in the realm of digital twins. A seamless integration and scalability in this context aren't just desirable traits—they're prerequisites for leveraging the full potential of digital twin technology. As digital twins bridge the physical and digital worlds, cloud computing enables them to flourish, offering the flexibility, processing power, and accessibility needed to handle complex operations.

When exploring the seamless integration of digital twins with cloud platforms, it's essential to consider the underlying infrastructure that supports such synergy. Cloud computing allows digital twins to tap into vast computational resources almost instantaneously, irrespective of geographical barriers. This elasticity ensures that as the need for processing power or storage capacity increases, cloud resources can be scaled up efficiently. This real-time adaptability not only optimizes performance but also minimizes costs, as organizations can adjust their resource consumption based on actual needs rather than static predictions.

Furthermore, cloud platforms provide a robust framework for integration with various technologies, ensuring that digital twins can connect with Internet of Things (IoT) devices, artificial intelligence (AI) systems, and other essential tools effortlessly. This interoperability is crucial for digital twins to function as real-time replicas of complex systems. Cloud environments support a myriad of protocols and standards, enabling seamless communication between different components, thus enhancing the accuracy and reliability of digital twins.

Scalability, on the other hand, assures that digital twin models can grow in complexity and size along with an organization's requirements. Today, a single digital twin might oversee an individual component or process. But tomorrow, it could be an entire city or a vast industrial network, which only cloud computing can support efficiently. Scalability isn't merely about growth in data handling capacity; it also involves expanding analytical capabilities to extract actionable insights from increasingly large and diverse data sets.

Cloud providers offer a plethora of services tailored for scaling digital twin applications. High-performance computing, data storage solutions like Data Lakes and Warehouses, and analytics tools such as big data processing engines, provide the foundation for scaling digital twin systems. With these tools, organizations can rapidly iterate and adapt their digital twin implementations, iterate simulations, and deploy updates, ensuring that the systems continually deliver value as they evolve.

An often overlooked aspect of cloud-driven scalability is global access. Cloud computing empowers teams distributed across the globe to collaborate in real-time on the same digital twin models. Such accessibility transcends geographical limitations, fostering innovation through diverse, global perspectives. This collaborative environment not only enhances problem-solving but also accelerates the development of cutting-edge solutions by incorporating insights from a variety of stakeholders.

In addition, leveraging cloud computing for digital twins paves the way for improved disaster recovery and data redundancy. By storing redundant copies of all critical data across different geographical locations, cloud solutions provide a safety net for organizations, ensuring business continuity even in the face of catastrophic failures. This reliability enhances trust and confidence in digital twin systems, making them a more attractive investment for risk-averse decision-makers.

Security, a primary concern in any cloud-based deployment, has seen significant advancements, making cloud environments increasingly secure. Providers implement rigorous security protocols, real-time threat detection, and comprehensive data protection strategies that ensure the integrity and confidentiality of data managed by digital twins. These measures make cloud-based digital twins not only scalable and integrative but also secure against ever-evolving cyber threats.

Although the advantages are manifold, the transition to cloud-based digital twins presents challenges that require careful navigation. Organizations must assess their existing IT infrastructure, address integration complexities, and consider regulatory constraints, particularly when dealing with sensitive data. Nonetheless, the continuous evolution of cloud computing and the introduction of hybrid solutions help mitigate these challenges, offering a balanced approach to leveraging the cloud's full potential.

The journey towards seamless integration and scalability isn't merely a technological shift—it's a transformational change in thought and operation. By embracing cloud computing, industries can unlock new realms of possibility with digital twins, driving innovation and competitive advantage. Organizations poised to explore these opportunities will navigate this dynamic landscape not just with current objectives in mind, but also with a strategic vision for what digital twins and cloud computing can achieve together in the future.

Ultimately, seamless integration and scalability in the context of digital twins are about creating systems that aren't just robust and responsive, but also transformative and forward-thinking. The future is one where digital twins are ubiquitous, and cloud computing will be the engine driving their expansion and refinement. As industries continue to embrace these innovations, the boundaries between the physical and digital will progressively blur, leaving a legacy of enhanced efficiency, creativity, and understanding in their wake.

Managing Data Overload

As we dive into the transformative role of cloud computing in digital twin technology, a pressing challenge emerges—managing data overload. In an era where data drives decisions, growth, and innovation, the amount and variety of data being generated are both a boon and a challenge. Digital twins, replicas of physical entities in a virtual space, thrive on vast streams of data. However, effectively managing this deluge requires innovative approaches and robust cloud solutions.

The interplay between digital twins and data is intricate. As these virtual replicas simulate and analyze real-world conditions, they produce a constant flow of information requiring storage, processing, and interpretation. The sheer volume of data can be staggering, especially when you're dealing with complex systems like smart cities or integrated healthcare networks. This is where cloud computing steps in, offering the scalability and flexibility to handle such immense data loads. Without cloud infrastructure, managing the data generated by digital twins would be nearly untenable.

Cloud platforms provide a seamless environment for processing and storing large datasets, enabling real-time analytics and insights. Traditional data storage solutions often fall short when dealing with the pace and volume of modern data flows. In contrast, cloud computing offers dynamic scalability, which is crucial for situations where data workloads can fluctuate dramatically. With cloud-based systems, it's possible to increase capacity on-demand, ensuring that data processing and storage remain uninterrupted. This dynamic scalability is essential in ensuring that digital twins continue to operate efficiently and effectively.

Beyond mere storage, the cloud provides advanced tools and services that facilitate data management and analysis. From machine learning to real-time analytics, these tools enhance the capabilities of digital twins, allowing them to make more accurate predictions, perform simulations, and offer insights. This ability to harness cloud-based solutions for analytics ensures that organizations can leverage the full potential of their data—turning raw information into actionable insights.

However, managing data overload isn't just about quantity; it's also about quality. The cloud offers advanced data management features that help ensure data integrity and reliability. This is pivotal for digital twins, where accurate and reliable data is the backbone of every simulation and prediction. Features like data validation, duplication, and error-checking enhance the overall trustworthiness of data into the digital twin ecosystem. With cloud platforms, organizations can implement robust data governance frameworks, ensuring that data is clean, consistent, and ready for analysis.

Security in managing data overload is another prime concern. With vast amounts of sensitive data moving to the cloud, protecting this information is paramount. Cloud service providers invest heavily in security measures, offering robust encryption, access controls, and compliance tools to safeguard data. For digital twins, which often handle sensitive information, these security features are non-negotiable. They ensure that while data is vast and flowing, it remains secure from breaches and unauthorized access.

It's also crucial to consider the impact of real-time data processing capabilities offered by the cloud. Digital twins require instantaneous data processing to function optimally, particularly in applications like real-time monitoring or predictive maintenance. The cloud's ability to process data in real-time ensures that digital twins can promptly react to changes, run simulations, and provide insights. This immediacy is key to the value digital twins bring to industries—transforming data into insights precisely when needed.

The cloud also fosters innovation in data management by offering platforms for integration. APIs and other cloud services facilitate seamless integration between different data sources, systems, and applications. This means that digital twins can draw data from diverse sources, unifying it within a cohesive model. The power to integrate data from multiple systems enriches the digital twin's ability to simulate real-world environments accurately, offering organizations an incredibly powerful tool for analysis and innovation.

Despite these capabilities, managing data overload requires strategic foresight and planning. Organizations need to develop comprehensive data management strategies, considering both current needs and future growth. Leveraging cloud services effectively calls for skilled professionals who understand both the technical and strategic aspects of cloud-based data management. Training and developing such talents must become an integral component of any organization embracing digital twins at scale.

Furthermore, collaboration between IT and business units is critical in managing data overload effectively. IT teams bring technical expertise in cloud environments and data management, while business units provide the context and application realm for digital twins. This interdisciplinary collaboration ensures that technical solutions align with business goals, platform needs, and organizational strategies. Together, they can devise solutions that maximize the potential of digital twins and the valuable data they generate.

Finally, the societal and ethical dimensions of managing data overload in digital twins cannot be overlooked. As data continues to burgeon, ethical considerations around privacy and use become increasingly significant. Cloud computing, while enabling vast data processing capabilities, also demands vigilant oversight and ethical standards. Adopting ethical data practices is a shared responsibility among entities deploying digital twins, ensuring fair and responsible data usage across industries.

In conclusion, managing data overload in the realm of digital twins is both a challenge and an opportunity. With cloud computing at the forefront, organizations can effectively harness, process, and secure massive data streams, enabling digital twins to deliver unprecedented insights and innovations. Success in this arena requires a delicate balance between robust technological solutions, strategic planning, and ethical data management. As we embrace the immense potential of digital twins, the role of cloud computing in managing data overload will only grow in importance, positioning it as a pivotal player in the digital revolution.

Chapter 15: Augmented and Virtual Reality Synergies

The synergy between augmented reality (AR) and virtual reality (VR) with digital twins unveils a frontier of interaction that feels almost magical. It brings digital twins to life, allowing users to engage in environments that blend the digital and physical worlds seamlessly. This confluence provides a rich, immersive experience, enhancing how we visualize and engage with complex datasets. Whether it's for design prototyping, infrastructure monitoring, or sophisticated training modules, AR and VR create vivid environments where ideas can be tested, refined, and realized with astonishing clarity. They transform static models into dynamic, interactive ecosystems, making abstract concepts tangible and insights more actionable. By leveraging this synergy, we empower industries to not only anticipate challenges but also innovate solutions in real-time, sparking a new era of efficiency and creativity.

Enhancing User Interaction

The marriage of augmented and virtual reality (AR and VR) with digital twins creates a powerful synergy that enhances user interaction like never before. This fusion is reshaping how users engage with complex systems, making interactive experiences more intuitive and immersive. At the heart of this transformation is the ability of AR and VR to deliver layered information in real time, enhancing user understanding and decision-making. By projecting digital twins onto the physical world or allowing users to experience them in fully virtual environments, these technologies offer a depth of interaction that bridges the gap between digital and physical realms.

Imagine a manufacturing plant where augmented reality displays real-time data from a digital twin overlaid on machinery. Maintenance workers can interact with this data without needing to sift through manuals or tap into computers. Instead, they can visualize thermal readings, wear and tear insights, and even diagnostic alerts right in front of their eyes. This translation of complex data into easily digestible visual cues accelerates understanding and problem-solving, ultimately reducing downtime and improving operational efficiency.

In the automotive industry, the integration of AR and VR with digital twins allows designers and engineers to both visualize and interact with car components before physical prototypes are even built. This not only speeds up the design process but also fosters innovation by permitting experimentation in a risk-free virtual environment. Seeing a car's digital twin animated in VR to simulate aerodynamics or mechanical performance provides unprecedented clarity over traditional CAD models or simulations.

The retail sector, too, benefits immensely from enhanced user interaction offered by AR and VR synergies. Digital twins of products can be visualized in a customer's living space through AR, allowing them to interact with the product before making a purchase. This interaction reduces return rates and enhances customer satisfaction since buyers get a realistic sense of how products fit or function in their environment before making decisions.

In healthcare, the combination of digital twins with AR and VR transforms training and patient care. Medical students can interact with human anatomy like never before, exploring digital twins of organs and systems through VR simulations. This hands-on experience deepens their comprehension and prepares them more effectively for real-world scenarios. Moreover, surgeons can use AR overlays during operations, guiding them with insights from digital twins, which enhances precision and patient outcomes.

Urban planners and architects also leverage AR and VR to visualize the development of digital twin cities. Stakeholders can walk through virtual streets and assess real-time data related to traffic flow, energy consumption, and air quality. This interactive engagement with digital twins aids in better decision-making regarding urban development and sustainable planning.

Educational institutions are fast embracing AR and VR-enhanced digital twin interaction to offer immersive learning environments. Students can engage with historical events, complex scientific concepts, or global ecosystems through simulated experiences that enhance understanding and retention. The interactive nature of these experiences anchors learning in a way traditional methods cannot.

Augmented and virtual reality are more than just display technologies; they become catalysts for deeper interaction with digital twins by adding a layer of contextual understanding. As users interact with digital twins, they are no longer passive observers but active participants in the digital narrative. This engagement fosters a deeper connection with the data and propels insight-driven actions.

Beyond visualization, AR and VR enhance tactile interaction, offering users haptic feedback as they engage with digital twins. This tactile feedback can be crucial in many industries, enhancing realism and providing nuanced control that mirrors physical interactions, thereby increasing user confidence and precision in tasks.

Furthermore, the synergy between AR/VR and digital twins fosters collaboration across geographical boundaries. Teams dispersed globally can interact with a digital twin of a product in real time, discuss modifications, and make informed decisions quickly. This collaborative potency reduces project timelines and costs while encouraging diverse input from multicultural teams.

While the technological advancements are thrilling, the focus now shifts to creating seamless user experiences that marry functionality with user-friendliness. This means designing intuitive interfaces that simplify user navigation and interaction with digital twins. Cognitive load must be minimized to enhance adoption rates and user satisfaction.

The immersive nature of AR and VR encourages a more personalized interaction with digital twins, tailoring experiences to the user's needs and context. For instance, in energy management, users can interact with a digital twin of their building to monitor energy consumption and receive personalized insights on optimizing usage, thereby promoting sustainable practices.

Finally, this dynamic fusion emphasizes the importance of feedback loops between users and systems. As users interact with digital twins through AR and VR, they generate data that enhances the digital twin's accuracy and utility. This iterative loop of user engagement and digital twin refinement is central to maximizing the potential of these technologies.

In summary, the synergy between augmented and virtual reality with digital twins is a game-changer in enhancing user interaction across a multitude of sectors. As these technologies continue to evolve, they will unlock even more innovative possibilities for user engagement, driving more profound organizational and societal transformations. Engaging with digital twins through AR and VR not only makes data more accessible but transforms it into a living, breathing entity that facilitates better decisions, efficiency, and creative breakthroughs. The potential is vast, and the future of user interaction looks incredibly promising.

Training and Simulation Applications

In the realm of augmented and virtual reality (AR and VR), one of the most transformative applications is undeniably in training and simulation. Imagine a world where complex scenarios can be enacted, analyzed, and mastered without the constraints of physical resources. This is where digital twins, combined with AR and VR, shine brightest. They open doors to immersive experiences and precision training, previously impossible to achieve.

Training in high-risk environments such as aviation, defense, and medicine has traditionally been fraught with difficulties ranging from safety concerns to cost limitations. Using AR and VR synergies, we can create a digital twin of any environment. This allows trainees to practice intricate procedures or respond to emergencies in a controlled, risk-free setting. Picture surgeons rehearsing delicate procedures with a high level of realism, or pilots honing their skills in a virtual cockpit that mirrors real-world dynamics. The potential to reduce error and enhance skill acquisition is staggering.

Let's delve into the specifics of how these technologies enhance training. AR overlays digital information onto the physical world, aiding users in performing tasks by providing real-time data and guidance. This technology is perfect for learning situations where understanding spatial relationships is crucial. Meanwhile, VR offers full immersion, enabling users to feel as if they're in an entirely different environment, perfect for simulations that require full engagement and detail orientation. Together, they create a synergy that's invaluable to modern training paradigms.

The customization possibilities in training and simulation using digital twins are virtually limitless. Tailoring scenarios to fit learner requirements or adjusting the difficulty levels on-the-fly allows for adaptive training that meets various learning paces and styles. Simulation platforms can record performance data, providing insights into user interaction and proficiency. This data can then be used to refine training programs, ensuring they remain relevant and effective. Furthermore, these capabilities extend beyond initial training to continuous education and skill enhancement, a critical consideration in industries characterized by rapid technological advancement.

For organizations, the economic benefits of using digital twins in AR and VR training are significant. The cost of setting up and maintaining physical training infrastructures can be exorbitant. By reducing dependencies on physical spaces and materials, and replacing them with virtual constructs, companies see a return on investment not just in terms of cost savings, but also improved workforce efficiency and reduced training times.

In the medical field, digital twins allow for patient-specific simulations. By creating a digital twin of a patient, medical professionals can simulate different treatment approaches, rehearsing surgeries in VR before performing the actual procedures. This practice underlines precision and personalized care, potentially improving patient outcomes while minimizing risks.

The gamification of training through AR and VR is another fascinating development. Incorporating game mechanics makes learning more engaging and increases motivation among participants. These immersive environments can simulate competitive scenarios, where learners earn rewards and achievements, further enhancing the training experience. By transforming serious training into an engaging activity, retention rates improve, and knowledge transfer becomes more effective.

Collaboration is yet another area enhanced by these technologies. Imagine teams from various geographical locations coming together in a virtual space that mirrors the real-world environment they are working on. Engineers, designers, and end-users can interact with a digital twin in real-time, discussing modifications and improvements as they see them unfold, leading to more effective communication and collaboration.

As the use cases for AR and VR in training and simulations expand, challenges such as user comfort, ease of use, and technology adoption emerge. However, with continuous refinement of hardware and software capabilities, much of this is being addressed. Modern advancements mean VR headsets are becoming lighter, more ergonomic, and less obtrusive. The integration of artificial intelligence can further customize and enhance the simulation experience, making it more intuitive and user-centric.

In conclusion, the amalgamation of digital twins with AR and VR for training and simulation applications promises a future of unlimited learning potential. As digital advancements progress, the simulation of more complex and intricate systems becomes feasible, facilitating the preparation and training of individuals for roles that require high precision and skill. It challenges traditional methodologies, paving the way for innovative, efficient, and effective training solutions. This convergence not only is an inspiration for those involved in technological education but also sets a precedent for future training methodologies across industries. Investing in these technologies is not just a step into the future, but a leap toward impressive improvements in how we train the workforce of tomorrow.

Chapter 16: Leveraging Machine Learning

As we delve deeper into the realm of digital twins, the integration of machine learning emerges as a powerful catalyst for innovation. Machine learning enables digital twins to transform static data into dynamic insights, making intelligent data analysis and predictive maintenance not only possible but remarkably efficient. By harnessing vast amounts of information, digital twins can anticipate failures, optimize operations, and support decision-making processes in unprecedented ways. This capability not only enhances the responsiveness and accuracy of simulations but also fuels a proactive approach to managing systems and resources. In this digital revolution, the synergy between machine learning and digital twins is redefining industries, inspiring organizations to adopt solutions that are both cutting-edge and pragmatically innovative. As you explore this intersection, consider the limitless possibilities it presents, from refining product designs to revolutionizing entire sectors with a focus on data-driven foresight.

Intelligent Data Analysis

As we delve deeper into the world of digital twins, machine learning emerges as a pivotal catalyst, enhancing the capability of these virtual models through intelligent data analysis. The data encompassed within digital twins isn't merely a collection of figures and metrics; it represents the digital pulse of their real-world counterparts. To decode this intricate heartbeat, intelligent data analysis offers the necessary tools and methodologies. By leveraging machine learning, we're not just analyzing data—we're learning from it, revealing insights that can drive innovation across industries.

Machine learning, a subset of artificial intelligence, transforms how digital twins process information. It thrives on patterns, extracting meaningful insights from vast data troves. Imagine manufacturing systems where digital twins of machinery predict potential breakdowns by noticing subtle changes in operation, or healthcare models identifying early warnings in patient health trends. Through intelligent data analysis, such feats are not just possible; they're increasingly becoming standard practice.

Incorporating machine learning into digital twins begins by gathering data from a range of sources, including IoT devices, historical databases, and real-time inputs. The challenge isn't just the collection, it's the curation. Intelligent algorithms sift through these data streams, narrowing down the needle-in-a-haystack moments that matter. And the power here is not in the sheer volume of the data but in the quality and context provided by sophisticated analysis techniques. Digital twins learn to adapt and respond to new data, evolving alongside the physical entities they represent.

One compelling aspect of intelligent data analysis is anomaly detection, a process that detects outliers in data that could indicate issues or inefficiencies. By continuously observing and learning from operational data, digital twins can autonomously flag potential problems—sometimes before they manifest in the physical world. This level of preemptive insight reduces downtime in manufacturing, ensures better reliability in automated vehicles, and enhances patient safety in healthcare. In essence, it transforms reactive maintenance into predictive maintenance, saving time, costs, and even lives.

Moreover, as machine learning algorithms refine their understanding, they can generate predictive models that guide future decision-making. These models consider historical data, identify trends, and propose potential outcomes, helping businesses strategize effectively. Whether it's forecasting energy consumption for smarter resource allocation or anticipating agricultural yields, intelligent data analysis equips stakeholders with a more informed perspective, enabling proactive rather than reactive strategies.

The integration of natural language processing (NLP) within digital twins marks another leap in data interaction. NLP allows systems to interpret and make sense of human language data, broadening the avenues through which insights can be extracted. For instance, sentiment analysis can assess customer feedback or employee satisfaction, offering a more nuanced understanding of qualitative data that can drive product improvement or employee engagement strategies.

The journey towards harnessing the full potential of intelligent data analysis is filled with challenges, but the rewards are substantial. Clear data governance frameworks are necessary to ensure the integrity and accuracy of data inputs. These frameworks establish guidelines for data collection, storage, and processing, building a foundation of trust and reliability. Emphasizing transparency in data practices also addresses ethical considerations, ensuring that analysis processes are not only robust but also ethically sound.

As with any use of advanced technologies, there are risks to mitigate. Data privacy and security are paramount, especially when dealing with sensitive information. Implementing secure architectures and adopting encryption methodologies are critical steps in safeguarding data integrity and confidentiality. Furthermore, regular audits and compliance checks fortify trust and ensure adherence to industry standards and regulations.

The cross-discipline collaboration that intelligent data analysis fosters is another dimension of its impact. Professionals from data science, software engineering, domain-specific experts, and technical architects collaborate, blending insights to refine digital twin models. This collaboration cultivates an ecosystem of innovation, where ideas converge to build solutions that are as versatile as they are effective.

Industries that have embraced intelligent data analysis in their digital twin strategies witness transformative outcomes. From smarter cities that dynamically adjust resources to meet population demands, to aerospace engineering pushing the boundaries of design possibilities, the ripple effects are being felt across sectors. The transition from data-rich environments to data-driven results is bridging the gap between potential and reality.

Looking ahead, the evolution of intelligent data analysis presents exciting horizons for digital twins. Increased computational power and the advent of quantum computing promise even more sophisticated pattern recognition and forecasting capabilities. As machine learning models grow in complexity and capability, the insights they yield will further empower decision-makers, leading to an era of unprecedented precision and agility in operations.

Ultimately, it's a process of visualization, optimization, and realization—a virtuous cycle fueled by data that's intelligently refined and purposefully applied. With each iteration, digital twins become not just mirrors of their physical counterparts but dynamic entities capable of anticipating and adapting to change. Embracing this power means opening the door to innovation, where intelligent data analysis is the key to unlocking a truly digital future.

Predictive Maintenance and Decision Support

In an era where operational efficiency and reliability are paramount, digital twins emerge as game-changers for predictive maintenance and decision support. At the core of this transformation lies the power of machine learning, enabling a proactive approach to maintaining assets and infrastructure. By leveraging data-driven insights, organizations can anticipate potential failures, optimize maintenance schedules, and make informed decisions, ultimately saving time and reducing costs.

Predictive maintenance traditionally relied on scheduled check-ups and historical failure data, a model often fraught with inefficiencies. Enter digital twins—virtual replicas that enable real-time monitoring and predictive analysis. By continuously collecting data from physical counterparts through sensors and IoT devices, digital twins provide a comprehensive picture of an asset's health. This continuous stream of information allows machine learning algorithms to detect patterns and anomalies indicative of degradation or impending malfunctions.

Consider the implications for a manufacturing plant. Instead of waiting for a machine to fail, digital twins can forecast its maintenance needs by analyzing wear and tear through real-time data. This foresight is not only about minimizing downtime but also about enhancing the longevity and performance of critical equipment. It's a shift from reactive repairs to a strategic maintenance regimen, ensuring assets are operationally fit at all times.

The benefits of predictive maintenance via digital twins extend beyond simple cost avoidance. The enhanced visibility into asset conditions facilitates better resource allocation and inventory management. Knowing precisely when a part will need replacement allows for just-in-time procurement, effectively streamlining inventory processes. This level of precision is pivotal in industries where timing and resource optimization can make or break profitability.

In addition to maintenance, digital twins excel in decision support. The integration of machine learning with digital twins equips decision-makers with a robust toolkit for evaluating complex systems. Whether it's optimizing a supply chain, adjusting to unforeseen disruptions, or planning for future expansions, the insights provided by digital twins are invaluable. As algorithms parse through vast datasets, they suggest actionable insights, backed by quantifiable evidence, about the most efficient courses of action.

One of the standout features of digital twins in decision support is their ability to simulate and analyze "what-if" scenarios. This capability is crucial in environments where decisions come with significant repercussions. For example, in energy plants, altering operational parameters can have widespread effects. Digital twins allow operators to run simulations, evaluating potential outcomes without affecting the physical system. The result is a more resilient and informed decision-making process.

Moreover, the predictive capabilities of digital twins are redefining risk management. By identifying vulnerabilities before they manifest into critical failures, organizations can mitigate risks effectively. This preemptive approach to risk is particularly beneficial in sectors like aerospace and defense, where safety is paramount, and the stakes are high. Through continuous learning from historical data and operating conditions, digital twins improve not only reliability but safety margins as well.

The integration of artificial intelligence and machine learning into digital twin platforms enhances their decision support functionalities. Advanced algorithms enable predictive insights that aren't just limited to maintenance—these can extend to operational enhancements and strategic innovation. For instance, digital twins optimized through AI can suggest operational changes that lead to energy savings, improved workflow efficiencies, or simplified logistics.

A critical aspect of deploying digital twins for predictive maintenance and decision support is the accessibility of data. With the influx of data from myriad sources, ensuring that the information is not only accurate but also relevant and actionable is essential. Here, cloud computing comes into play, providing scalable storage solutions that enable the seamless integration and processing of vast amounts of data, fortifying the digital twin's analytical prowess.

The journey toward adopting digital twins isn't solely about technological integration; it demands a shift in mindset. Organizations must be prepared to embrace a culture of continuous improvement and data-driven decision-making. Training and development initiatives become essential to equip the workforce with the necessary skills to leverage these advanced technologies effectively. Cross-disciplinary collaboration will be pivotal, merging insights from data scientists with practical knowledge from industry specialists.

Looking ahead, the potential for digital twins in predictive maintenance and decision support is immense. As technologies evolve and machine learning models become more sophisticated, the accuracy and scope of predictions will only enhance. Industries will continue to benefit from improved uptime and operational efficiencies, ultimately contributing to an organization's competitive edge in a data-driven world.

In conclusion, digital twins offer a transformative approach to maintenance and decision-making, bridging the digital and physical worlds. They represent a new frontier where data is king, and informed action becomes the norm. With digital twins, industries across the spectrum can anticipate the future, respond proactively, and thrive amidst challenges, ensuring they not only keep pace with technological advancements but lead with it.

Chapter 17: Business Models and Economic Impact

The journey toward harnessing digital twins for tangible economic benefits begins with crafting business models that are not just innovative, but also adaptable to dynamic market conditions. By integrating digital twin technologies into existing operations, businesses can unlock new revenue streams and drive efficiency gains across various sectors. Consider the monetization of data insights generated by digital twins; organizations can leverage these to offer advanced, value-added services that enhance customer engagement and foster brand loyalty. The shift provides a competitive edge, enabling companies to tap into emerging market trends while staying agile. Digital twins redefine industry landscapes, from predictive analytics offering maintenance solutions to real-time simulations optimizing resource allocation. Each example underscores a critical economic impact: reduced operational costs and improved profitability. As industries evolve, the adoption of digital twins becomes a strategic necessity, empowering organizations to innovate responsibly, adapt swiftly, and contribute meaningfully to economic growth.

Monetizing Digital Twin Technologies

The potential for monetizing digital twin technologies lies not only in their capability to replicate physical environments virtually but also in the myriad applications they unlock across various industries. As advancements in technology persist, the economic landscape continuously evolves, offering new and exciting opportunities to capitalize on digital twins. Recognizing how to transform these virtual replicas into tangible value forms a crucial aspect of modern business strategies.

One of the primary avenues for monetizing digital twins is through the enhancement of product lifecycle management (PLM). By integrating digital twins into PLM, companies can achieve a substantial reduction in costs associated with product development and testing. Companies can simulate and refine their designs virtually, minimizing the need for physical prototypes, which are often expensive and time-consuming to produce. This ability not only speeds up the time-to-market but also allows firms to innovate more rapidly, gaining a competitive edge in their respective industries.

Digital twins empower businesses to offer advanced predictive maintenance services, a transformative approach that can significantly extend the lifespan of machinery and equipment. By utilizing the real-time data and insights that digital twins provide, companies can predict equipment failures before they occur, optimizing maintenance schedules and reducing unplanned downtime. This proactive maintenance strategy appeals to industries heavily reliant on equipment reliability, such as manufacturing and energy sectors, and represents a lucrative opportunity for service-based revenue models.

Moreover, digital twins facilitate the creation of new service offerings. By leveraging the data and insights gathered from these virtual models, businesses can develop innovative solutions tailored to customer needs. For example, a manufacturer could offer personalized optimization services to clients, adjusting machinery settings based on digital twin simulations to improve efficiency and output. This customization not only adds value to clients but also allows companies to differentiate their offerings, thereby increasing revenue potential.

In the realm of smart cities, digital twins enable urban planners and developers to optimize city infrastructures and services efficiently. By simulating various scenarios, governments and private entities can identify cost-saving measures, enhance resource allocation, and streamline operations. These improvements in urban efficiency translate into substantial financial savings and open the door to new business models centered around consultancy and urban optimization contracts.

The agriculture sector also stands to benefit immensely from digital twin technologies. With precision farming techniques supported by digital twins, farmers can make data-driven decisions that optimize crop yield and resource usage. By offering subscription-based access to such insights and analytics, companies can generate recurring revenue streams while helping farmers increase their productivity and profitability.

Additionally, digital twins can transform supply chain management by providing end-to-end visibility and enhancing decision-making processes. This transparency allows companies to streamline operations, reduce waste, and effectively manage inventory, leading to significant cost savings. These efficiencies can be translated into competitive pricing advantages or enhanced value propositions for customers, further driving economic gains.

The technology sector itself benefits from developing and deploying digital twin solutions. From software developers creating the underlying platforms to consultancy firms advising on integration and deployment, an entire ecosystem of businesses thrives around digital twin technologies. As more enterprises adopt and leverage digital twins, these supporting industries experience growth, contributing to the broader economy.

However, monetizing digital twin technologies necessitates an understanding of the diverse market demands and technological landscapes. Businesses must tailor their strategies to align with the specific needs of their target sectors, ensuring they can capitalize on the unique opportunities digital twins provide. This strategic alignment requires collaboration among various stakeholders, including technological experts, business leaders, and end-users, to fully realize the economic potential of digital twins.

Digital twin technologies do more than replicate physical assets; they create avenues for innovation, efficiency, and new business models. By embracing these virtual replicas, businesses can unlock substantial economic value, translating technological advancements into market success. As industries continue to evolve and digital twins become more sophisticated, the possibilities for monetization are limited only by the imagination and ambition of those who wield them.

Competitive Advantages and Market Trends

In the rapidly evolving landscape of digital twins, the competitive advantages are manifold and touch virtually every industry. Companies that adopt digital twin technologies gain a significant edge by enhancing operational efficiency, reducing costs, and improving product quality. At the heart of these benefits lies the ability of digital twins to simulate real-world conditions, allowing businesses to test hypotheses, predict outcomes, and make data-driven decisions. This capability transforms not just business processes but the industries at large, making organizations agile and more responsive to market demands.

A key competitive advantage of digital twins is their ability to deliver substantial cost savings. By creating a virtual replica of a physical asset or process, companies can preemptively identify potential issues and optimize performance without the disruptive and costly implications of real-world trial and error. Maintenance, for example, becomes predictive rather than reactive, significantly decreasing downtime and extending lifespan of machinery. Industries ranging from manufacturing to energy are seeing reduced operational expenses and improved reliability, leading to a positive impact on the bottom line.

The precision and accuracy of digital twins play a crucial role in product development cycles. With the capacity to simulate complex systems and processes, businesses can experiment with design configurations and various scenarios to maximize performance before a single physical prototype is produced. This accelerates innovation at a scale that was previously unimaginable, opening up new opportunities for personalizing products and services, which in turn can boost market share and customer loyalty.

Additionally, digital twins are instrumental in enhancing sustainability and resource efficiency, a growing priority across all sectors. Aligned with global pushes for environmental responsibility, this capability allows organizations to better manage resources, reduce waste, and meet stringent compliance and regulatory standards. Companies are increasingly leveraging digital twins to track and improve their ecological footprint, which not only enhances their brand image but also satisfies increasingly eco-conscious consumers and stakeholders.

The market trends surrounding digital twins suggest a robust and growing demand. According to industry reports, the digital twin market size is poised for exponential growth, predicted to reach multibillion-dollar valuations in the next few years. This growth is fueled by advances in IoT, AI, and big data analytics, which provide the necessary infrastructure to deploy digital twins at scale and sophistication. Businesses across industries are investing heavily in digital twin capabilities, indicating a profound shift in digital and operational strategies.

Moreover, as digital twin technology matures, the barriers to entry decrease, making it accessible for small and medium-sized enterprises (SMEs) to leverage the benefits traditionally available only to larger organizations. This democratization bolsters market competition while fostering innovation and technology adoption across the board. SMEs are particularly finding digital twins valuable for their potential to level the playing field against larger competitors through strategic technology integration.

Another exciting trend is the integration of digital twins with other transformative technologies. When paired with the power of cloud computing, digital twins can scale efficiently, managing unprecedented amounts of data to enhance decision-making processes. Similarly, Augmented Reality (AR) and Virtual Reality (VR) technologies offer new dimensions for interacting with digital twins, enabling more intuitive and immersive user experiences. These synergies are opening new pathways in industries ranging from healthcare, where patient outcomes can be modeled more accurately, to retail, where consumer experiences can be tailored with remarkable precision.

Turning towards the future, the continued evolution of digital twin technology promises even greater impacts. The convergence of emerging technologies like 5G and edge computing is paving the way for real-time, high-fidelity digital twins that offer deeper, more instantaneous insights. As these capabilities expand, digital twins will not only predict but also prescribe solutions, further enhancing their role in strategic decision-making processes. The consequential acceleration in innovation will likely spur more collaborative ventures and strategic partnerships, both within and across sectors.

In summary, the competitive advantages and emerging market trends underscore the transformative potential of digital twins. Organizations incorporating these advanced solutions are not merely optimizing existing processes but reimagining them altogether, unlocking value and setting new benchmarks in service delivery and operational excellence. As such, digital twins are not just a tool of the future; they are a pivotal component of the present and a catalyst for revolutionary advancements across the global economy.

Chapter 18: Regulatory and Compliance Aspects

In the rapidly evolving domain of digital twins, navigating the web of regulatory and compliance landscapes presents both challenges and opportunities. As industries increasingly rely on digital twin technology, adherence to established standards and practices becomes crucial to ensure safety, security, and interoperability across sectoral boundaries. The interplay between innovation and regulation demands that companies remain agile and informed, integrating compliance strategies seamlessly into digital architectures. This proactive approach not only mitigates legal risks but also fosters a culture of trust and accountability. Meanwhile, the formulation of new regulations is essential to accommodate the dynamic nature of digital twins, encouraging sustainable growth and ethical development. For technology enthusiasts and decision-makers, understanding these regulatory frameworks is key to leveraging digital twins as a transformative tool for innovation and progress.

Standards and Best Practices

As we journey deeper into the world of digital twins, the significance of standards and best practices in ensuring successful implementation cannot be overstated. These standards form the bedrock on which interoperable and secure digital twin systems are built. The orchestration of digital twins across diverse sectors necessitates a unified language and methodology—a goal to which various international bodies and industry consortiums are committed.

At the heart of these efforts is the development of open standards. The term "open standards" refers to specifications that are publicly available and are designed to provide a common technical foundation for the systems they address. These standards are crucial for promoting interoperability, allowing different digital twin systems to communicate and function together seamlessly. Interoperability isn't just a technical requirement; it's a catalyst for innovation. By breaking down the silos between systems, open standards foster an environment where ideas can flow freely, paving the way for breakthroughs that might have otherwise been stifled by proprietary boundaries.

Organizations like the International Organization for Standardization (ISO) and the Institute of Electrical and Electronics Engineers (IEEE) play pivotal roles in the development and dissemination of these standards. Their work is indispensable in crafting guidelines that ensure digital twins are designed, implemented, and operated within a consistent framework that practitioners across industries can adopt. The impact of their contributions can be seen in areas such as data exchange protocols, security frameworks, and even the operational benchmarks that define what a "successful" digital twin should achieve.

Alongside formal standards, best practices are the experiential knowledge accrued over time, often through trial and error, that provide practical guidance for leveraging digital twins effectively. These best practices are shared by pioneers and practitioners who have navigated the complexities of digital twin deployments and have found strategies that yield the best results. Emphasizing the role of continuous learning and adaptation, sharing best practices allows for knowledge transfer across projects and industries, helping others to avoid pitfalls and replicate successes.

Best practices in data governance, for instance, highlight the need for establishing robust frameworks to manage data integrity, security, and privacy. Data is the lifeblood of digital twins, and managing it effectively ensures that digital twins function as insightful, reliable proxies for their physical counterparts. This involves adhering to protocols that prioritize data accuracy, protect against cyber threats, and comply with regulatory requirements like GDPR or HIPAA, depending on the sector in question.

Moreover, the evolution of best practices stresses the importance of stakeholder collaboration throughout the digital twin lifecycle. From ideation and design through to implementation and maintenance, engaging diverse stakeholders—from engineers to end-users—ensures that the digital twin meets the collective needs and expectations. Oftentimes, innovation emerges when diverse perspectives come together, challenging assumptions and expanding the realm of possibilities.

The convergence of standards and best practices isn't isolated to risk mitigation—it's a proactive strategy to bolster the resilience and scalability of digital twins. Incorporating real-time data streams, predictive analytics, and feedback loops can enhance the adaptability of digital twin systems to changing environments and conditions. Feedback loops, in particular, serve as a best practice by continuously refining and improving system performance based on insights gleaned from digital twin operations.

The dynamic nature of the digital landscape requires a commitment to adaptability. Standards and best practices aren't static documents but living artifacts that evolve in response to technological advancements and shifting market demands. As such, there is an imperative for organizations to stay abreast of changes in the standards landscape and to contribute to the dialogue, ensuring that emerging standards reflect the practical realities faced by practitioners at the coalface.

In essence, standards and best practices provide the skeleton and muscles for digital twin implementations. They don't just help ensure compliance and mitigate risk; they empower organizations to innovate with confidence. With a structured approach, firms can align their digital twin strategies with broader technological goals, unlocking new opportunities for efficiency and growth.

As the digital twin ecosystem continues to grow, embracing standards and best practices fosters a more open, collaborative, and innovative future. By adhering to these principles, organizations not only ensure the success of their own initiatives but contribute to shaping a more integrated digital world. That's the promise of digital twins—a promise realized through commitment to standards and best practices that guide us all toward a brighter, more connected future.

Navigating Legal Landscapes

In the rapidly evolving domain of digital twins, understanding the legal terrain is imperative. As digital twins bridge the realm between the physical and digital, navigating the associated legal landscapes demands careful consideration and proactive strategies. These technologies are not only reshaping industries but also challenging existing regulatory frameworks. Companies eager to harness the potential of digital twins must appreciate and anticipate the multifaceted legal challenges.

It's essential to recognize that the legal implications of digital twins vary across jurisdictions. For technology enthusiasts and industry professionals, the importance of strategic legal foresight can't be overstated. As digital twins transform sectors such as healthcare, manufacturing, and retail, they intentionally or inadvertently push the boundaries of current regulations. This necessitates a thorough understanding of compliance requirements in different regions.

The integration of digital twins into traditional systems not only opens up dynamic possibilities but also introduces complex legal questions. Particularly in sectors like healthcare, digital twins must align with stringent regulations concerning patient data privacy and security. For example, compliance with laws such as the Health Insurance Portability and Accountability Act (HIPAA) in the United States is critical. Companies need robust legal frameworks to manage the sensitive interplay between digital representations and real-world entities.

Arguably, data privacy is one of the most pressing legal considerations. Digital twins generate and process vast quantities of data, often including personal and sensitive information. This level of data handling necessitates compliance with data protection regulations like the General Data Protection Regulation (GDPR) in Europe. Businesses must embed privacy-centric design principles into their digital twin technologies to comply with such regulations, ensuring that user permissions and data confidentiality remain paramount.

Furthermore, there's the vital aspect of intellectual property (IP) rights. With digital twins, the line between what is physical and digital can become blurred. This confusion raises questions about ownership and rights over digital models, especially when these models incorporate data from multiple sources or entities. Legal teams have to grapple with ensuring that IP laws address the nuanced realities of digital twins, thus protecting companies' innovations and proprietary information from infringement.

In certain industries, digital twins can play a crucial role in regulatory compliance itself, by offering advanced simulation capabilities for testing and validation. In aerospace and defense, for example, digital twins can be utilized to simulate extreme conditions, helping companies to meet the rigorous compliance standards for safety and performance. These simulations allow for detailed documentation and evidence required during regulatory audits or certifications.

Moreover, the potential export and international application of digital twins introduce concerns about cross-border data transfers. Companies often operate in a global environment, where data protection laws and compliance measures differ substantially from one country to another. To successfully navigate these landscapes, businesses must implement robust data governance frameworks and leverage legal expertise to address the complexities of international regulations.

The advent of digital twins also prompts the need for updated contracts and service level agreements (SLAs). As businesses leverage digital twins for predictive analytics and operational efficiencies, contracts must clearly define liability, data ownership, and usage rights. These documents must be meticulously crafted to protect parties involved while maximizing the potential of digital twin deployments.

As digital technology advances, ethical considerations and compliance with emerging standards become inseparable from legal landscapes. Stakeholders must assess the societal impact of digital twins, ensuring that innovations do not inadvertently result in discrimination or bias. Incorporating ethics into the regulatory and legal framework will foster trust and responsibility across the industry.

Finally, industry associations and regulatory bodies are evolving their standards to accommodate the unique challenges posed by digital twins. Engaging with these bodies can provide businesses with valuable insights into best practices and forthcoming regulatory changes. Active participation in shaping these standards can not only facilitate compliance but also influence the direction in which the industry evolves.

Navigating the legal landscapes of digital twins is undeniably challenging, but it is also an opportunity to build resilient, adaptive, and legally sound systems. By committing to a path of compliance and innovation, organizations can unlock the transformative potential of digital twins while safeguarding their operations from legal pitfalls. The journey demands vigilance, adaptability, and a collaborative spirit to align technological advancements with regulatory imperatives.

Chapter 19: Case Studies of Success

Across various industries, digital twins have emerged as transformative tools, driving innovation and creating new benchmarks for success. In manufacturing, digital twins have enhanced production lines, ensuring optimal efficiency and reducing downtime through predictive maintenance. Healthcare has witnessed groundbreaking applications, with patient-specific models enabling personalized treatment plans and advancing medical research. Smart cities leverage digital twins for sophisticated urban planning, improving public services and infrastructure resilience. The automotive sector utilizes these virtual counterparts to elevate vehicle performance and safety standards. In energy, digital twins optimize power plant operations and facilitate the integration of renewable sources, advancing toward a more sustainable future. These case studies underscore the potential of digital twins to revolutionize industries by providing valuable insights, fostering innovation, and ultimately, transforming how we approach complex challenges. The lessons learned offer a roadmap for best practices, guiding organizations to harness the full potential of digital twin technology.

Notable Implementations in Various Sectors

As we delve deeper into the realm of digital twins, it's crucial to highlight some of the stand-out implementations across diverse sectors. These examples not only showcase the versatility of digital twin technology but also serve as a blueprint for future innovations. By examining these implementations, we can appreciate the broad spectrum of opportunities that digital twins present to industries worldwide.

The aviation industry, for instance, has been at the forefront of leveraging digital twin technology. In aircraft design and testing, companies like Boeing and Airbus utilize digital twins to simulate the entire lifecycle of an aircraft. This approach allows engineers to identify potential issues long before physical prototypes are constructed, saving both time and resources. Furthermore, digital twins empower engineers to simulate extreme conditions, such as high-altitude weather, ensuring that the final product meets rigorous safety standards.

In manufacturing, General Electric (GE) has demonstrated significant success with digital twin applications. By incorporating digital twins into their operations, GE has achieved enhanced production efficiency and predictive maintenance capabilities. Their power plants employ digital twins to monitor equipment health in real time, predicting failures before they occur, thus reducing downtime and maintenance costs.

The healthcare sector has also benefited enormously from digital twin technology. Hospitals and research institutions use patient-specific digital twins to revolutionize personalized medicine. Imagine a virtual replica of an individual's anatomy, where treatments can be tested for efficacy and side effects in a risk-free environment. This application helps tailor medical treatments to individual needs, significantly improving patient outcomes.

Notably, smart cities are integrating digital twins to improve urban planning and infrastructure management. City planners use virtual models to predict traffic patterns, optimize public transport, and manage utilities efficiently. By employing digital twins, cities can simulate the effects of new policies, allowing for data-driven decisions that can lead to smoother, safer urban environments.

In the automotive sector, companies like Tesla are at the cutting edge of digital twin applications, using them to enhance vehicle performance and safety. With every Tesla car effectively functioning as a rolling digital twin, real-time data is continuously fed back to improve design and function. This constant feedback loop not only enhances the driving experience but also anticipates potential mechanical failures, ensuring timely interventions.

Meanwhile, the energy sector is transforming its operations through digital twin adoption. Siemens, a leader in this field, uses digital twins to optimize their power generation systems. This results in more efficient energy production and distribution, reducing waste and contributing to sustainability goals. By simulating different energy scenarios, companies can adapt quickly to the dynamic energy landscape, including the integration of renewable resources.

Construction and real estate industries are reimagining project management with digital twins. By creating virtual replicas of buildings and infrastructure, stakeholders can collaborate more effectively throughout the project lifecycle. Construction firms can visualize the stages of development, identify potential issues, and streamline the entire building process. This not only reduces costs but also ensures that projects are completed on time and to specification.

The retail industry is harnessing digital twin technologies to gain deeper insights into consumer behavior. Companies like Walmart use digital twins to analyze shopping patterns, optimize store layouts, and personalize the shopping experience. By understanding customer preferences in a virtual environment, retailers can enhance product offerings and improve customer satisfaction, ultimately driving sales growth.

In the realm of telecommunications, digital twins are becoming integral to network infrastructure planning. Companies like Verizon and AT&T use them to model and simulate network performance, ensuring optimal coverage and capacity. By employing digital twins, these companies can predict network congestion and adapt in real-time to maintain service quality for their customers.

Moreover, agriculture is a sector where digital twins are growing in influence, particularly in precision farming. By using digital twins to model crop growth and soil conditions, farmers can make informed decisions about irrigation, fertilization, and pest control. This technology helps optimize resource usage, increase yields, and promote sustainable farming practices.

These notable implementations highlight how digital twins are not merely a theoretical concept but a transformative force capable of reshaping numerous sectors. The ability to create a dynamic, virtual representation of physical entities offers unprecedented opportunities. As more industries embrace this technology, the innovations stemming from its application will likely redefine our understanding of efficiency, sustainability, and growth.

Lessons Learned and Best Practices

Much like any other technological journey, the path to successful implementation and utilization of digital twins is paved with both triumphs and challenges. These case studies provide invaluable insights into the lessons learned along the way and establish a set of best practices that can be adopted by organizations seeking to harness the transformative power of digital twins.

A recurring theme among successful digital twin implementations is the critical importance of a clear and well-defined objective from the outset. Organizations that have excelled typically begin with a thorough understanding of their specific needs and problems that digital twins can solve. This clarity ensures that the technology is deployed effectively, aligning with strategic goals and yielding tangible benefits. It's a lesson that emphasizes the need to start with the end in mind while remaining flexible to adapt as the technology and business landscape evolve.

Moreover, collaboration plays a pivotal role in the successful deployment of digital twins. The collaborative synergy between multiple stakeholders—including developers, engineers, data scientists, and end-users—often determines the project's success. It's not enough to simply acquire the technology; continuous communication, feedback loops, and iterative improvements are essential components of a thriving digital twin ecosystem. Effective collaboration ensures that digital twins are fully integrated into the operational fabric and meet the diverse needs of all stakeholders.

Another vital takeaway from successful digital twin case studies is the significance of data integrity and security. With digital twins heavily reliant on data inputs to mirror their physical counterparts, ensuring data accuracy and maintaining confidentiality is paramount. Organizations need robust data governance frameworks and cybersecurity measures to protect sensitive information. Lessons learned here show that neglecting these areas can lead to vulnerabilities, circumventing the potential gains from digital twins.

Investing in skill development and organizational change management has also proven crucial. As digital twins reshape traditional workflows and decision-making processes, there is often a need for upskilling employees to maximize their potential. Many successful implementations have been accompanied by comprehensive training programs that empower teams to excel in a digital twin-driven environment. This proactive approach not only mitigates resistance to change but also fosters a culture of innovation and continuous improvement.

Flexibility and adaptability come up repeatedly as characteristics of organizations that leverage digital twins effectively. Industries are evolving rapidly, and technological advancements can quickly shift the landscape. Successful implementations often stem from a willingness to pivot strategies as needed. This iterative learning process helps organizations stay ahead and maximizes the impact of digital twins on their operations and outcomes.

Furthermore, beginning with smaller, pilot projects often impacts overall success. It allows organizations to test theories, refine models, and assess impacts without significant upfront investment or risk. Pilots provide a safe environment to learn and improve, ensuring the lessons carried over to full-scale deployments increase the likelihood of success.

There is another critical lesson in recognizing and managing the potential ethical implications associated with digital twin technologies. As these tools become more integral, ethical considerations surrounding data privacy, consent, and transparency need to be thoroughly vetted and addressed. Successful case studies highlight the importance of establishing ethical guidelines and frameworks early in the process to maintain trust and integrity within the digital twin landscape.

Finally, there is much to be said about leadership and vision. Visionary leaders who understand the potential of digital twins and advocate for their strategic use often catalyze transformations that ripple throughout their organizations. Leadership that fosters an environment where innovation is encouraged, and calculated risks are supported, leads to more enthusiastic adoption and deeper buy-in from the workforce.

In summary, the lessons learned from successful digital twin implementations underscore the multidimensional nature of adopting this technology. These include setting clear objectives, fostering collaboration, ensuring data security, upskilling the workforce, embracing flexibility, running pilot tests, addressing ethical considerations, and cultivating visionary leadership. By drawing from these experiences, organizations can better navigate the complexities of digital twin adoption and unlock unprecedented opportunities for innovation and growth.

Chapter 20: Future Trends and Innovations

As we stand on the brink of technological evolution, the future of digital twins is charged with potential that's limited only by our imagination. Emerging trends suggest a symbiotic integration of artificial intelligence, edge computing, and next-gen IoT devices, creating hyper-realistic simulations that transcend today's capabilities. These innovations promise to transform industries from healthcare to urban development, offering unprecedented precision and efficiency. Imagine digital twins evolving to predict system failures before they occur, thus revolutionizing maintenance schedules and significantly reducing downtimes. As the frontier of digital twins advances, businesses and societies will push for broader applications, encouraging a deeper fusion of the physical and digital worlds. This transformative journey will inspire industry pioneers to reimagine the boundaries of what's possible, setting the stage for innovations that could redefine our interaction with technology in daily life.

Emerging Technologies and Applications

Digital twins are becoming a key innovation driver across multiple industries, thanks to converging technologies such as artificial intelligence (AI), the Internet of Things (IoT), and advanced data analytics. As these technologies evolve, digital twins move from initial hype to practical applications, offering unprecedented opportunities for innovation and optimization. The fusion of these technologies creates a dynamic landscape where digital twins not only mirror real-world entities but also enhance decision-making processes, predict future scenarios, and revolutionize business models.

One groundbreaking trend is the integration of digital twins with AI capabilities. AI enhances digital twins by processing vast amounts of data to learn from patterns and optimize outcomes. For instance, in complex manufacturing systems, AI-driven digital twins can simulate production lines and identify potential bottlenecks or inefficiencies, providing actionable insights that lead to cost savings and productivity improvements. This synergy between AI and digital twins opens possibilities for self-optimizing systems that adapt over time without human intervention.

Another transformative application is the incorporation of edge computing. By processing data closer to the source, edge computing reduces latency and bandwidth use, which is crucial for real-time analytics and monitoring. In smart cities, edge-enabled digital twins can manage and analyze data from connected sensors scattered throughout urban environments, delivering insights on energy consumption, traffic flow, and public safety in near real-time. This close-to-source processing capability makes digital twin applications faster and more reliable, offering a seamless experience even in data-intensive environments.

Augmented reality (AR) and virtual reality (VR) are also reshaping the potentials of digital twins. These immersive technologies provide intuitive and interactive visualization of complex data sets. In sectors like aerospace and defense, AR/VR-powered digital twins enable engineers and technicians to interact with three-dimensional models of aircraft systems, facilitating maintenance and repair tasks with high precision. Essentially, AR/VR bridges the gap between digital models and physical operations, enhancing human capabilities and simplifying complex procedures.

Quantum computing is an emerging frontier that's poised to impact digital twin capabilities significantly. Although still in its developmental stages, quantum computing promises an exponential increase in processing power, offering the ability to solve complex simulations and optimization problems much faster than classical computers. As quantum technology matures, digital twins could utilize this power to perform intricate simulations involving multiple variables and scenarios, such as predicting climate change impacts on urban environments or optimizing global supply chain networks.

The healthcare industry also stands on the cusp of transformation with digital twins fueled by the latest technological trends. By integrating digital twins with genomic data and advanced analytics, personalized medicine becomes feasible. Digital twins of patient physiology can track and predict disease progression, enabling tailored treatment strategies and improving outcomes. This personalized approach not only enhances patient care but also makes healthcare systems more efficient and predictive in nature.

Blockchain, known for its security and decentralization capabilities, is finding its place in digital twin ecosystems as well. In supply chain management, blockchain-secured twins can offer traceable and immutable records of product journeys, enhancing transparency and trust between stakeholders. This technology ensures the integrity and security of data in digital twins, which is critical when coordinating between multiple parties in complex networks.

As digital twin technologies advance, they present new business models and economic opportunities. Companies can leverage them for competitive advantage by offering digital twin-driven services, such as predictive maintenance solutions or personalized consumer experiences. As industries adopt these models, we see shifts towards subscription-based services and result-oriented offerings that reflect the increasing importance of continuous optimization and personalized solutions.

Furthermore, the ethical and regulatory landscape around digital twins continues to evolve. Organizations must navigate these considerations carefully while deploying digital twins effectively. Ensuring data privacy, security, and ethical usage isn't just a compliance challenge but a fundamental requirement for trust and adoption. Emerging frameworks and best practices will likely shape how digital twins are employed, balancing innovation with responsibility.

Cross-industry collaborations are vital for harnessing the full potential of emerging technologies in digital twins. Partnerships among technological innovators, industry experts, and regulatory bodies can spur the development of versatile solutions that address real-world challenges. Open-source platforms and collaborative ecosystems can expedite innovation, allowing shared knowledge and resources to uplift mastery and delivery of digital twin technology across sectors.

Ultimately, the trajectory of digital twins is driven by their capacity to integrate, simulate, and optimize in ways previously unimaginable. With emerging technologies continually reshaping their capabilities, digital twins are set to redefine how we understand, interact with, and improve the world around us. Embracing these advancements will empower organizations to unlock tremendous value, inspire creative solutions to longstanding problems, and champion a future where innovation remains at the forefront of all endeavors.

The Next Frontier for Digital Twins

The digital twin technology has captured the imagination of businesses and technologists alike, captivating us with promises of unprecedented insights and operational efficiencies. While current applications transform sectors like manufacturing, healthcare, and urban management, there's a broader horizon ripe with potential—the next frontier for digital twins. This chapter delves into new territories, exploring how digital twins are set to redefine our interactions with technology and the world.

As technology advances at a breakneck speed, one pivotal area where digital twins could make a significant impact is in the integration with advanced artificial intelligence and machine learning capabilities. Imagine a digital twin not just as a virtual replica but as an evolving entity, continuously learning and adapting. This dynamic interaction isn't merely about replicating physical entities but understanding them in a way that allows for predictive modeling and proactive solutions. The integration of AI can enable digital twins to forecast potential issues, optimize processes autonomously, and offer insights that were previously unattainable.

Consider the potential of digital twins in environmental sustainability. As the world grapples with climate change, a technology that can accurately model ecological impacts before implementation could be a game-changer. Digital twins could simulate complex environmental processes, helping scientists and policymakers test and refine sustainable practices. This can lead to more informed decision-making on issues like climate mitigation, conservation efforts, and resource management. The frontier here is not just technological innovation but a commitment to planetary stewardship.

Another compelling application beckons within the realm of personalized healthcare. The future could witness an individual digital twin for every human, a sophisticated replica that merges genomic data, lifestyle habits, and medical history. Such a model could propel us into a new era of personalized medicine, where treatments are tailor-made based on the predictive analysis of these digital replicas. The predictive power of digital twins could also revolutionize drug development by identifying the best molecular compositions and dosage schedules tailored to individual biochemistry, substantially reducing time and costs associated with traditional clinical trials.

The relationship between digital twins and augmented reality (AR) is another frontier poised for exploration. Innovators are already experimenting with the concept of overlaying digital twins onto real-world views through AR devices. This blending of digital and physical worlds holds exciting potential for sectors ranging from construction to education. For example, architects can walk inside a building's digital twin before construction begins, exploring designs in real-time and making adjustments before the first brick is laid. In education, virtual classrooms embedded with interactive digital twins could offer an immersive learning experience, tailoring educational content to each student's pace and style.

Further accentuating the forward march of digital twins are advancements in edge computing. Edge computing enables data processing at or near the source of data generation rather than relying on a centralized cloud. This capability reduces latency and allows for real-time decision-making, addressing the limitations that come with data transmission and processing delays. The implication is profound for digital twins that require instant data feedback and interaction, such as in autonomous vehicles and smart city infrastructure management.

A crucial aspect of this next frontier is the ethical framework that will govern the deployment of digital twins. As these systems become more deeply integrated into daily life, questions of privacy, data ownership, and ethical use become paramount. The governance of these digital siblings will demand vigilant oversight to prevent misuse and ensure that the benefits of digital twins are distributed equitably across society. Industry leaders, policymakers, and ethicists must collaborate to devise regulations that uphold transparency while fostering innovation. Addressing these ethical questions head-on will be critical in gaining public trust and broader acceptance of digital twins.

Incorporating robust cybersecurity measures into digital twin platforms will also be essential. As the scale and complexity of digital twin applications grow, so do the risks associated with data breaches and cyber threats. The sensitivity of information contained within digital twins necessitates an innovative approach to security, one that integrates advanced encryption, anomaly detection, and rapid response strategies. Cyber resilience will be a key focus area as digital twins become more prevalent in mission-critical operations.

Industry collaboration and the fostering of open-source digital twin platforms could be another driver in navigating this new frontier. Collaborative ecosystems that promote shared learning and innovation will propel the development of new applications while helping to standardize practices across industries. Open-source initiatives can lower barriers to entry, inviting diverse solutions and accelerating the adoption of digital twins globally. It will be crucial for businesses and governmental entities to support and invest in such collaborative ventures to unleash the full potential of this technology.

Beyond the traditional industrial applications, digital twins can become a catalyst for creativity and expression. Artists and storytellers are beginning to tap into the potential of digital twins to craft next-level interactive experiences. In the entertainment industry and digital art, digital twins could offer personalized immersive experiences, where narratives change based on individual interactions. This integration presents a wholly new dimension of user engagement, turning consumers into co-creators and audiences into participants.

As we stand at the cusp of this next frontier for digital twins, the imperative is clear: strategize for a paradigm shift where digital twins are integral to everything from daily routines to the grand challenges humanity faces. This journey won't be without challenges; however, those who embrace digital twins' full potential and navigate its complexities will position themselves at the heart of tomorrow's innovations. The horizon isn't just about achieving technological marvels but also about envisioning a future woven intricately with possibility and promise.

Chapter 21: Building a Digital Twin Strategy

Creating a robust digital twin strategy isn't just about adopting new technology; it's about envisioning a transformative future for your organization. You need a strategic plan that aligns with your vision and encompasses detailed roadmap development while ensuring adaptable organizational change management. The journey involves understanding the unique challenges and opportunities digital twins bring to your industry. A successful strategy requires the integration of cross-functional expertise, harnessing data insights, and fostering innovative thinking throughout your team. By doing this, organizations can anticipate market shifts and stay ahead of the curve, ultimately transforming their operational dynamics. Embrace change, cultivate a culture of continuous learning, and lead with a vision driven by data, tailored to sustain value in an ever-evolving digital landscape.

Strategic Planning and Roadmap Development

The journey toward a successful digital twin strategy begins with clear and strategic planning. It's not merely about adopting new technology; it's about creating a structured framework that aligns with your organization's goals. To start, a comprehensive understanding of current processes and systems is vital. This ensures that the digital twin initiative complements and enhances existing operations, rather than creating redundancy or conflict. The strategy must pivot on both the specific needs of the industry and the unique capabilities of digital twins to transform traditional practices.

At the heart of strategic planning is the necessity of identifying clear objectives. What exactly should the digital twin achieve for your organization? Whether it's improving operational efficiency, enhancing customer engagement, or driving innovation, each goal must be well-defined. From there, measurable outcomes are crucial. With defined metrics in hand, organizations can track progress and ensure that the digital twin is delivering the desired value.

The planning should take an iterative approach, allowing for adjustments as the strategy unfolds and matures. Flexibility is essential given the rapid advancements in technology. A static plan could quickly become obsolete. Hence, integrating feedback loops is crucial. These loops enable continuous refinement based on real-world performance and evolving business needs.

Technology adaption is a key consideration. It involves an in-depth analysis of current IT infrastructure, alongside the integration capabilities of various digital twin components. It's important to assess whether the existing systems can support the new demands of data processing and storage, or if upgrades will be necessary. Equally, the security protocols must encompass the new technology landscapes and ensure compliance with regulatory standards. One must not overlook the scalability; the digital twin should grow with the enterprise, adapting to increased demand or new applications as needed.

Furthermore, a well-designed roadmap should articulate the timeline of implementation phases. By breaking the project into smaller, manageable phases, organizations can address challenges progressively. Short-term, mid-term, and long-term goals help maintain focus while minimizing risk. Each phase should be designed to build on the success of the previous one, creating momentum and quick wins that demonstrate value to stakeholders. This phased delivery also supports resource allocation, ensuring that the necessary human and capital resources are secured at the right times.

Another critical component of strategic planning is stakeholder engagement. Involving stakeholders from various levels of the organization fosters alignment and commitment. Their insights and feedback enrich the strategy, enhancing its relevance and effectiveness. Engaging with external partners, such as technology providers and consultants, can also offer invaluable expertise and perspectives that strengthen the strategic approach.

A robust digital twin strategy should also address potential challenges upfront. Mapping out risks and planning for mitigation strategies can prevent disruptions. Discussing potential pitfalls and preparing for them, whether they're related to technological limitations, cultural resistance, or market dynamics, is indispensable. Proactively addressing these issues can help maintain the momentum and confidence in the strategy's trajectory.

Training and development are also cornerstones of strategic planning. As digital twins are integrated, building the necessary skill sets within the workforce ensures effective implementation and utilization. This might involve training programs, workshops, or even partnerships with educational institutions to develop talent with specialized knowledge. The workforce must feel equipped and confident in their abilities to adapt to and leverage digital twin technologies.

Finally, the role of leadership cannot be overstated. Leaders must champion the digital twin strategy, displaying commitment and vision. Their support, advocacy, and clear communication are pivotal in fostering a culture of innovation and change. By articulating the benefits and sharing successes, leaders inspire confidence and enthusiasm across the organization, paving the way for a successful digital twin transformation.

Organizational Change Management

Organizational change management is a pivotal aspect in building a digital twin strategy. It serves as the bridge between groundbreaking technology and its smooth adoption within an organization. Embracing digital twins isn't just about the technology. It's about reshaping mindsets, redefining processes, and realigning organizational cultures to support and sustain digital transformation. To achieve this, organizations must be prepared to navigate through a complex landscape of human behavior, entrenched practices, and potential resistance.

When introducing digital twins, a thorough understanding of existing organizational practices is crucial. Companies must assess their current state and the cultural environment they operate in. This knowledge helps in identifying potential hurdles that may arise during implementation. Every organization has its own unique characteristics and dynamics, which is why a one-size-fits-all approach to change management simply won't work. Customizing strategies to match the organizational culture can greatly enhance acceptance and integration of digital twins.

Communication plays a central role in organizational change management. It's not just about informing employees of upcoming technological shifts but engaging them in the journey. People need to understand not only what changes are happening but also why they're necessary. Establishing a clear narrative that links the adoption of digital twins to organizational goals fosters a sense of purpose and direction. Open channels for feedback and discussion can mitigate resistance and promote a cooperative atmosphere.

Moreover, leadership must champion the change. It’s essential for leaders to exemplify the behaviors and attitudes they wish to see throughout the organization. They need to be visible, approachable, and proactive in addressing concerns or misconceptions about digital twins. This kind of leadership inspires confidence and minimizes uncertainty. Training programs and workshops can further empower employees, equipping them with the necessary skills and knowledge to thrive within a digitally driven ecosystem.

Getting buy-in from stakeholders across all levels is also critical. Each stakeholder group will have distinct perspectives and interests, ranging from operational concerns to strategic business impacts. Understanding and addressing these interests helps in crafting a comprehensive change management plan that takes into consideration the multifaceted nature of organizational ecosystems. Aligning digital twin initiatives with individual and departmental goals ensures broader participation and commitment.

It's equally important to identify change advocates within the organization. These are key individuals who can influence and mobilize their peers towards embracing new technologies. They can serve as intermediaries, translating the technical aspects of digital twins into practical, relatable terms. By leveraging these advocates, companies can foster a more inclusive and supportive learning environment for all employees.

Establishing measurable milestones throughout the change process can also contribute vastly to organizational adaptation. Clearly defined objectives allow for the monitoring of progress and the identification of areas requiring additional focus or resources. The feedback gathered from these assessments provides invaluable insights that inform continuous improvement efforts. When milestones are achieved, celebrating these successes can reinforce positive behaviors and enhance motivation.

Another crucial element is ensuring the alignment of digital twin strategies with existing business processes. Disruptive technologies like digital twins necessitate rethinking of traditional workflows to maximize their potential benefits. Redesigning these processes should be a collaborative effort, with input from those directly involved in their execution. This inclusive approach not only improves the efficiency and effectiveness of digital twin applications but also encourages ownership of the processes among employees.

In workforce management, safeguarding the organization's talent is essential. As digital twins require new skills and competencies, upskilling and reskilling initiatives become indispensable. By investing in continuous learning opportunities, organizations empower their employees to grow alongside the technology. Developing a workforce that is both knowledgeable and adaptable ensures long-term sustainability of digital twin strategies.

Finally, continuous assessment and flexibility in strategy implementation are vital. Organizations must be agile enough to adapt to unexpected challenges or opportunities that may arise during the digital twin journey. This agility is achieved through a culture of experimentation and learning where feedback loops are embraced and innovation is encouraged. By being open to modifying strategies based on real-world experiences, companies can stay ahead in the digital domain.

Chapter 22: Collaboration and Partnership Models

In the realm of digital twins, collaboration and partnership models serve as the catalysts for innovation, driving technological advancements and expanding possibilities. They enable a symbiotic engagement among diverse stakeholders, from industry experts to tech start-ups, aligning their strengths towards a common goal. This interconnected ecosystem fosters open-source platforms and shared developments, empowering organizations to pool resources and knowledge for groundbreaking discoveries. By nurturing a network of shared aspirations and expertise, these models invite a transformative shift where creativity flourishes and bold ideas are rapidly prototyped and deployed. Such dynamic partnerships not only accelerate technological adoption but also ensure that digital twin solutions are robust, scalable, and inclusive by design. Ultimately, embracing these collaborative frameworks can lay the groundwork for sustainable growth and a resilient digital future, where innovation is fueled by the collective vision and effort of many rather than the isolated achievements of a few.

Multi-Stakeholder Engagement

Digital twins are reshaping industries as we know them, promising revolutionary changes in how businesses and systems operate. To unlock their full potential, however, engaging multiple stakeholders is vital. This involves a complex interplay of various parties, each with its distinct interests, skills, and resources. Effective multi-stakeholder engagement can foster innovation, drive interoperability, and ensure sustainable implementation of digital twin technologies.

Multi-stakeholder engagement means more than just getting everyone together; it's about creating meaningful collaboration. This often involves diverse participants such as industry leaders, technologists, government entities, academia, and sometimes consumers themselves. Each stakeholder brings unique insights and capabilities. Industry leaders may provide an understanding of market needs and competitive landscapes, while technologists focus on the practicality and functionality of digital twins. Government entities can offer regulatory guidance and support, ensuring that implementations adhere to necessary standards and regulations.

One of the key challenges in multi-stakeholder engagement is aligning these varying interests toward a common goal. Stakeholders might have different, sometimes conflicting, priorities. Overcoming these differences requires transparent communication and collaboration strategies that focus on shared values and objectives. Partnerships can benefit from structured decision-making processes, where every stakeholder's voice is heard. Techniques such as consensus building and facilitated workshops can help navigate these complexities, bringing stakeholders together in a way that respects each viewpoint while fostering cooperation and innovation.

In the context of digital twins, the role of academia cannot be understated. Academic institutions often lead the way in research and development, providing the initial breakthroughs and theoretical foundations necessary for digital twin technologies. By engaging with academic institutions, industry players can stay on the cutting edge of technological advancements, benefitting from the fresh insights and critical thinking that academia provides.

Government and regulatory bodies play another crucial role in multi-stakeholder engagements for digital twins. With their influence and legislative power, they ensure that all developments adhere to legal and ethical standards. This can affect various aspects, from data privacy and security to environmental impacts. Their involvement not only adds legitimacy to digital twin projects but also helps in scaling the adoption across different sectors, creating a supportive environment for innovation.

Non-governmental organizations (NGOs) and public interest groups can also be indispensable partners in this ecosystem. They bring perspectives that focus on societal benefits and environmental considerations. Their advocacy and participation can guide the ethical implementation and utilization of digital twins, ensuring that these technologies contribute positively to broader societal goals. This also includes emphasizing sustainability and resource efficiency, which are becoming crucial factors in digital transformation initiatives.

A successful multi-stakeholder engagement strategy relies heavily on open communication channels. Digital platforms that facilitate easy interaction and data sharing among stakeholders are essential. These platforms can act as a nexus for ideas, allowing stakeholders to brainstorm and collaborate in real-time, regardless of geographic location. The use of these collaborative tools needs to be encouraged to sustain engagement over time and adapt to the ever-evolving nature of digital twin projects.

Trust is another fundamental component of multi-stakeholder engagement. Stakeholders must trust that their contributions will be valued and incorporated into the project's outcomes. This trust is built through transparency in operations and decision-making processes, as well as ensuring data integrity and security. Safeguarding intellectual property and respecting privacy are paramount to nurtur relationships among stakeholders, fostering a cooperative atmosphere where innovation can flourish.

Furthermore, to maximize the potential of digital twins through stakeholder engagement, there must be a focus on building and nurturing ecosystems rather than just individual partnerships. An ecosystem implies a network of interconnected relationships and interactions that can evolve and grow, adapting over time to new challenges and opportunities. Within these ecosystems, stakeholders should be encouraged to play to their strengths, contributing unique expertise and resources that complement those of others.

Training and capacity-building efforts also form an essential component of multi-stakeholder engagement. By investing in skill development and education, stakeholders can ensure that participants are well-equipped to contribute to digital twin projects. This is crucial in a rapidly advancing technological landscape, where new skills are continuously needed. Educational workshops, joint training exercises, and knowledge-sharing platforms not only enhance collaboration but also empower stakeholders to innovate.

Another approach to enhancing multi-stakeholder engagement is leveraging public-private partnerships (PPPs). PPPs can be particularly effective in digital twin deployments that require long-term commitments and substantial investments. By sharing risks and resources, PPPs can promote sustainable development and provide innovative solutions to challenges that are too complex for either sector to handle alone.

Ultimately, successful multi-stakeholder engagement in digital twin projects leads to enhanced outcomes that benefit all parties involved. It ensures that these cutting-edge technologies can be efficiently rolled out, scaled, and adapted to various sectors. The benefits of channels such as better urban planning, smarter manufacturing systems, or improved healthcare services can be widely shared. Engaging diverse stakeholders not only drives progress and safeguards interests but also enhances the collective capability to tackle future challenges.

Open Source and Collaborative Platforms

In the expansive universe of digital twins, collaboration is not a mere convenience—it's a necessity. One of the most potent tools in this collaborative arsenal is the open-source platform. By bringing together a multitude of perspectives and expertise, open-source platforms enable the sharing of knowledge, fostering innovation like never before. These collaborative environments encourage a diversity of ideas, speeding up the development of cutting-edge solutions that can drive the digital twin revolution.

Open source stands as a beacon of accessibility and collective problem-solving. It breaks down barriers by allowing anyone with an interest and a bit of technical know-how to contribute to and benefit from ongoing projects. This accessibility reduces development costs and accelerates innovation timelines. When individuals or companies contribute to open-source projects, they're not just sharing code—they're building a community of practice that spans industries and geographies. In the digital twin landscape, this can mean faster iterations of improvements and a more resilient technological framework.

Collaboration doesn't stop at code sharing. Many platforms also provide a venue for sharing use cases, methodologies, and even challenges. This transparency builds trust and encourages more stakeholders to get involved, which is particularly crucial in sectors like healthcare, where digital twins are becoming integral to patient-centric solutions. Here, open-source platforms enable medical professionals, technological innovators, and policy makers to pool their skills and knowledge, contributing to solutions that are both innovative and sensitive to patient needs.

However, the benefits of open-source and collaborative platforms aren't limited to individual applications or use cases—they extend to overarching architectural frameworks and infrastructure. By sharing foundational software components and tools, developers across industries can create interoperable systems. This interoperability is a cornerstone of digital twin ecosystems, allowing disparate systems to work together seamlessly. For example, a shared platform might enable a city's digital twin systems to interact effectively with the digital twins used in the transportation sector, enhancing urban planning and development.

Given the rapid pace of technological advancement, open-source platforms offer a crucial agility that proprietary models might lack. They enable continuous, organic, and community-driven evolution of technologies. This adaptability is especially valuable for the digital twin environment, which requires constant innovation and the ability to scale effectively as more data is accumulated and new applications are explored. Open source provides not just flexibility in development but also in deployment, allowing organizations to tailor tools to their specific needs without being locked into a rigid vendor roadmap.

One of the shining examples of open-source collaboration in digital twins is the evolving role of cloud-native platforms that offer shared environments for experimentation and deployment. Platforms like Kubernetes have set technical benchmarks by enabling digital twin applications to scale across the cloud efficiently. This scalability accelerates innovation by reducing the time and resources needed to deploy new ideas into testable models. It serves as a foundation upon which other technological advancements in artificial intelligence and machine learning—integral to digital twins—can be built cohesively.

Moreover, open-source and collaborative platforms cultivate a thriving environment for educational opportunities and workforce development. When professionals from various backgrounds and expertise levels can contribute and learn in an open-source space, the collective knowledge pool deepens. This is exceptionally advantageous for emerging markets or sectors just beginning to realize the potential of digital twins, providing them with the resources and community support needed to climb the learning curve effectively.

Open-source efforts also have the power to democratize technology. By providing accessible entry points into sophisticated technologies, they can help close the digital divide. Enthusiasts and smaller companies with fewer resources can access the same tools and opportunities as larger, well-funded corporations. This democratization can lead to more equitable competitive landscapes and push innovation boundaries in unexpected ways. Democratically structured platforms can drive technology adoption across diverse sectors, from agriculture to aerospace, bringing the benefits of digital twins to a broader audience.

The role of governance in open-source projects can't be underestimated either. Structured governance models ensure that open-source initiatives remain aligned with community goals while maintaining code quality and security standards. These governance frameworks are particularly important in the digital twin domain, where the stakes involve both complex technologies and critical real-world applications. Successful governance can lead to increased participation and trust in open-source platforms, paving the way for groundbreaking progress in digital twin technology.

While the prospects are incredibly promising, the path isn't free from challenges. One of the primary concerns in collaborative open-source platforms is the need for effective collaboration mechanisms. These mechanisms include setting common objectives, managing contributions, resolving conflicts, and maintaining momentum. With diverse participants contributing from worldwide locations, ensuring clear communication and shared understanding across diverse cultural contexts is essential.

As digital twins become more prevalent, open-source and collaborative platforms will only grow in importance. However, the question remains: How can we ensure that these platforms not only foster innovation but do so sustainably and responsibly? Addressing this requires a careful balance of innovation-driven openness with protective measures for data privacy and security, especially given the rising concerns detailed in other sections of this book.

Bringing it all together, the relationship between digital twins and open-source platforms is symbiotic. By fostering a culture of collaboration and openness, we enable technologies to cross boundaries they might otherwise encounter. Through shared efforts and a commitment to collective advancement, the potential to transform industries is not just a possibility—it's a guarantee. Partnerships of this nature remind us that collaboration is indeed the keystone of innovation, an idea uniformly powerful across the technology landscape, but especially so in the realm of digital twins.

Chapter 23: Skills and Knowledge for the Digital Twin Era

As we navigate the complexities of the digital twin era, the demand for a workforce adept in both traditional and emerging technologies has never been more pronounced. Developing expertise in data analytics, IoT systems, and AI is crucial, but it's equally important to harness interdisciplinary collaboration. Professionals must not only be versed in advanced technical skills but also possess the ability to communicate and innovate across domains. Organizations should foster environments where continuous learning and adaptability are valued, enabling employees to pivot as technologies evolve. The ability to synthesize complex information, coupled with strategic thinking, will empower decision-makers to harness the full potential of digital twins. This transformative era calls for a dynamic approach to education and training, one that equips stakeholders with the tools needed for both immediate challenges and future innovations. In embracing this paradigm shift, we set the stage for unprecedented opportunities and advancements across industries.

Workforce Development and Training

As the digital twin era rapidly evolves, the need for a workforce equipped with relevant skills and knowledge becomes imperative. The convergence of physical and digital realms in industries demands that professionals not only understand core digital twin concepts but also stay adaptable amidst continuous technological advancements. For organizations aiming to leverage digital twins effectively, workforce development and training are cornerstones.

At the heart of this transformation is the shift in required competencies. Traditional roles are expanding, and new roles are emerging, characterized by a blend of technical prowess and soft skills. Understanding how to maneuver within a digital twin ecosystem—comprising IoT, AI, machine learning, and simulation technologies—is now as critical as collaboration and problem-solving abilities. Thus, initiatives that foster these competencies are paramount.

Training programs need to evolve alongside digital twin technologies. A robust curriculum should not only cover theoretical underpinnings but emphasize hands-on, experiential learning. Immersive training using virtual and augmented reality can simulate real-world scenarios, offering learners a tangible sense of digital twin applications. This experiential approach ensures that theoretical knowledge is augmented by practical insights.

The corporate sector can play a pivotal role by investing in continuous professional development. Companies should encourage lifelong learning through internal training sessions, workshops, and partnerships with educational institutions. Industry conferences and seminars can further enhance this learning by providing platforms for knowledge exchange and networking, keeping professionals abreast of the latest innovations and trends.

Moreover, interdisciplinary collaboration should be a priority. Digital twins exemplify a multidisciplinary approach, interweaving fields such as data science, engineering, and business management. Facilitating collaboration across these disciplines can foster innovation and lead to groundbreaking solutions. Teams should include a diverse range of expertise to harness the full potential of digital twins, ensuring a holistic approach to problem-solving.

For technology enthusiasts and industry professionals eyeing the future, embracing a mindset geared towards adaptability and continuous improvement is crucial. The pace of change in digital twin technology is fast, and the tools and methods employed today may evolve within a short span. Thus, professionals should cultivate a mindset that welcomes change and seeks out learning opportunities proactively.

Decision-makers have their roles cut out too. They can shape organizational culture by embedding learning and development in the company ethos. By doing so, they create an environment where knowledge sharing and skill enhancement are not only encouraged but are part of the organization's identity. This cultural shift can lead to more agile organizations, capable of nimbly adapting to technological advancements in the digital twin space.

Furthermore, it's essential for training programs to be inclusive and accessible. As digital twin technologies become prevalent across various sectors, there's a significant opportunity to bridge skill gaps and foster diversity within tech-centric roles. Programs tailored to different levels of expertise and career stages can ensure that everyone, from newcomers to seasoned professionals, can contribute to and benefit from digital twin innovations.

Governments and educational institutions also have pivotal roles. By aligning educational curricula with industry needs, they can prepare future generations for careers in this field. Encouraging STEM education from an early age and integrating digital twin concepts into higher education programs is critical for cultivating a talent pool ready to tackle future challenges.

Public-private partnerships can further accelerate workforce development initiatives. These collaborations can facilitate the exchange of resources, expertise, and insights, creating synergies that benefit both sectors. Such partnerships can lead to more comprehensive training programs, increased funding for educational initiatives, and greater alignment with industry demands.

The digital twin era is not just about technology; it's about leveraging technology to unlock human potential. By placing workforce development and training at the forefront, we set the stage for a future where digital twins are not only tools for innovation but become integral to our professional and societal growth. In doing so, organizations and individuals alike can harness the transformative power of digital twins to drive progress and inspire change.

Interdisciplinary Collaboration

In an era defined by the convergence of digital and physical realms, digital twins stand at the crossroads of technological innovation and practical application. For digital twins to realize their full potential, interdisciplinary collaboration is not just beneficial—it is essential. The development and deployment of digital twin technology require the seamless integration of multiple fields, including engineering, computer science, data analytics, and business management, among others. Such collaboration sparks innovation by bringing varied expertise to the table, ultimately leading to breakthroughs that a single discipline might not achieve.

Consider the implications in the healthcare sector, where digital twins are helping to transform patient care. Assembling a team of experts in medical sciences, data analytics, and software development can significantly enhance patient-centric innovations. For instance, medical professionals might identify a specific patient need, which data scientists address using predictive analytics, while software developers create the necessary interfaces for implementation. This collaboration does more than create a digital representation of a patient; it enables real-time monitoring and predictive insights that can save lives.

The automotive industry provides another example where interdisciplinary collaboration thrives. Here, digital twins are employed to enhance vehicle performance and safety. Engineers, data scientists, and simulation experts work together, utilizing smart algorithms from data analytics, contextual insights from mechanical engineering, and predictive capabilities driven by machine learning. Their collective input ensures that the digital twins are accurate and reliable, thereby reducing costs and improving safety features.

Interdisciplinary collaboration isn't just advantageous; it's a necessity for smart cities, where digital twins help optimize urban planning and infrastructure management. Urban planners, civil engineers, and environmental scientists have to work together to create sophisticated models that simulate urban environments. The insights derived from these models can lead to more efficient public services, reduced environmental impact, and improved quality of urban life. Here, collaboration doesn't just bridge disciplines; it transforms them, creating synergies that might not have been possible otherwise.

Agriculture and food production are fields ripe for collaborative innovation. Precision farming benefits immensely from the combined efforts of agricultural scientists, environmental engineers, and software developers. By leveraging data from digital twins, farmers can make informed decisions about crop planting and resource allocation, thus improving yields and promoting sustainability. This kind of collaboration not only optimizes agricultural practices but also contributes to addressing global food security challenges.

The construction and real estate sectors also see vast improvements through interdisciplinary collaboration. When architects, civil engineers, and IT specialists converge, construction projects can benefit from digital twin models that optimize project management and enhance building performance. The collaboration allows for the creation of efficient structures that are not only built quicker but are more sustainable and adaptable to future needs.

Interdisciplinary collaboration hinges on mutual respect and the blending of diverse skill sets. One of the challenges often faced is the dialogue between experts who may not initially speak the same technical language. Bridging this communication gap requires a shared vision and understanding of the objectives pursued through digital twin technology.

The education and training sectors play a crucial role in facilitating this collaboration by designing curricula that are interdisciplinary in nature. As educational institutions shift towards more integrated learning approaches, they help prepare a workforce capable of navigating complex, multi-disciplinary projects. By equipping students with skills that span multiple disciplines, future professionals are better poised to thrive in environments that demand collaboration.

The synergy between academia and industry is another cornerstone of effective interdisciplinary collaboration. Research institutions and companies can partner to create digital twin solutions that are both innovative and practical. Collaborative research initiatives can lead to the development of cutting-edge applications that not only respond to current industry needs but also anticipate future challenges and opportunities.

Despite the potential challenges, the benefits of interdisciplinary collaboration in developing digital twins are immense. The diverse viewpoints and skills that collaborative teams offer lead to more robust and creative solutions. By fostering a culture of open communication and continuous learning, organizations can harness the full potential of digital twins.

In conclusion, the digital twin era demands a shift in how we approach collaboration. By embracing interdisciplinary methods, we create opportunities that extend far beyond traditional boundaries, inviting innovation that is both holistic and transformative. As digital twins continue to evolve, so too must our approach to working together across disciplines. Embracing this change will ensure that we not only keep pace with technological advancement but also lead it into a future full of promise.

Chapter 24: Measuring Success and Return on Investment

As we delve into the realm of digital twins, a crucial element emerges: evaluating their impact through clear metrics. Measuring success and return on investment (ROI) demands a keen understanding of key performance indicators that align with strategic objectives. It's not just about immediate financial gains; the long-term benefits and sustainability also define the essence of a successful digital twin implementation. The challenge lies in capturing both tangible and intangible outcomes, from enhanced operational efficiencies to innovative capacities that drive competitive advantages. By aligning metrics with goals, decision-makers can illuminate the path to substantial returns, ensuring that digital twins are not just a transient trend but a transformative force in the landscape of modern technologies.

Key Performance Indicators

In today's fast-paced technological landscape, determining the success and return on investment of digital twins can be a complex task. Key performance indicators (KPIs) provide a vital framework for evaluating these investments. As digital twins become integral across various sectors, understanding which KPIs to focus on can enable decision-makers to discern not only the value but the potential to innovate further.

First, consider operational efficiency. Digital twins are lauded for their ability to simulate and optimize processes. Therefore, one crucial KPI is the reduction in operational costs. By comparing baseline figures before the implementation of digital twins with post-implementation data, organizations can benchmark their improvements. The insights gained from predictive maintenance and real-time monitoring often translate into significant savings, identifying this as a pivotal area for KPI assessment.

Another important KPI is process cycle time. With digital twins, processes can be continuously improved through simulation and iteration. The reduction in the cycle time of production or service delivery not merely highlights efficiency gains but also underlines the agility of the organization in adapting to market needs. Cheers to reduced latency and optimized workflows—hallmarks of a successful digital twin deployment!

The quality of outputs, products, or services is yet another KPI where digital twins play a transformative role. The ability to simulate scenarios enables testing without the usual constraints of physical prototypes, which often leads to fewer defects and higher customer satisfaction. By tracking defect rates and feedback before and after digital twin adoption, decision-makers can make informed evaluations of enhanced quality metrics.

Effective risk management is an often understated yet profound KPI. For sectors where precision and safety are paramount, like aerospace or healthcare, digital twins offer unparalleled predictive capabilities. By simulating risk and addressing potential failures proactively, organizations not only protect resources but enhance strategic resilience. Risk minimization through predictive insights in a dynamic environment proves how integral this technology can be in ensuring operational continuity.

In addition to performance-focused metrics, customer satisfaction and experience provide critical insights into the success of digital twin initiatives. With real-time data analysis and customer behavior modeling, digital twins can improve user experiences across interfaces and touchpoints. Metrics such as customer retention rates and net promoter scores can be leveraged to assess the impact these technologies have on end-user engagement.

Resource utilization is another crucial KPI related to environmental impact. As sustainability becomes a competitive differentiator, digital twins allow enterprises to optimize resource usage. Whether it's reducing energy consumption in smart grids or improving material utilization in construction, resource efficiency metrics provide clear guidelines on sustainability objectives being met. Observing these parameters becomes instrumental in aligning with corporate responsibility goals.

The strategic value of digital twins also reflects on a company's innovation capabilities. A fundamental KPI in this regard is the time-to-market for new products or services. Digital twins facilitate faster iterations and validation cycles, often bridging the gap between conception and delivery. Companies that expedite their time-to-market reap the benefits of competitive positioning and faster revenue recognition.

Moreover, data-driven decision-making is a central advantage of digital twins. The KPI here revolves around the accuracy and quality of decision markers derived from vast reservoirs of data. The degree to which data insights influence executive decisions and strategic pivots can be monitored through solution-derived ROI figures and measurable business outcomes.

The exploration of these KPIs leads to long-term sustainability in organizational growth. By consistently evaluating and refining processes through the lens of these indicators, businesses not only justify their initial investments but fold those learnings into ongoing innovation frameworks. Through intentional KPI alignment, digital twins prove to be more than just technological enablers; they shape company futures.

As we dive deeper into the age of digital twins, the necessity of well-defined KPIs cannot be overstated. These performance indicators serve as a compass, guiding businesses toward efficient, intelligent, and sustainable advancements. By strategically focusing on the most relevant KPIs, organizations can continue to create immense value, driving both current success and future potential.

Long-Term Benefits and Sustainability

In a rapidly evolving technological landscape, digital twins stand out as a pivotal innovation that extends beyond immediate gains, offering profound long-term benefits. At the heart of this enduring impact is the ability of digital twins to provide continuous, real-time insights and improvements across various sectors. By creating a virtual counterpart to physical entities, organizations can better anticipate challenges, streamline operations, and minimize waste—an approach that not only enhances efficiency but also contributes to a more sustainable future.

The long-term benefits of digital twins are deeply intertwined with sustainability goals. One of the primary ways digital twins facilitate sustainability is by enhancing resource management. For instance, in industries such as manufacturing and agriculture, digital twins enable precise monitoring and control of resources, reducing unnecessary consumption and promoting more efficient use of materials. This strategic optimization not only cuts costs but also aligns with global efforts to reduce carbon footprints and promote environmental stewardship.

Moreover, digital twins drive sustainability by enabling predictive maintenance, which extends the lifespan of machinery and infrastructure. This proactive approach prevents breakdowns and reduces the need for frequent replacements, minimizing the depletion of raw materials. In turn, organizations can significantly decrease their environmental impact while maintaining steady operational flow, an outcome that resonates with both economic and ecological priorities.

Beyond the environmental benefits, digital twins also offer substantial socioeconomic advantages. As organizations adopt and refine digital twin technologies, they naturally progress towards more adaptable and resilient business models. The ability to simulate scenarios and evaluate potential outcomes allows for more informed decision-making, reducing risks and enhancing profitability over time. This adaptability is crucial in an era where market dynamics are unpredictable and consumer expectations are continuously shifting.

Investment in digital twin technology also plays a critical role in fostering innovation. With ongoing advancements in artificial intelligence, machine learning, and the Internet of Things, the potential applications for digital twins are expanding at an unprecedented rate. These integrations not only improve the accuracy and functionality of digital twins but also encourage continuous research and development efforts. By prioritizing innovation, organizations stay ahead of the curve, ensuring long-term viability and competitiveness in their respective markets.

Furthermore, digital twins empower workforce development by necessitating a new skill set for employees. As industries evolve, the demand for expertise in digital twin technology, data analysis, and related fields grows, prompting educational institutions and organizations to prioritize upskilling and reskilling initiatives. This focus on building a knowledgeable workforce supports long-term economic growth and job creation, further highlighting the societal benefits of digital twins.

The sustainable integration of digital twins necessitates collaboration across various stakeholders. By fostering partnerships between businesses, governments, and academic institutions, the development and deployment of digital twins can be better streamlined and regulated. Such collaborations pave the way for the creation of shared knowledge bases and best practices, ensuring that the benefits of digital twins are equitably distributed across sectors and communities.

In conclusion, the long-term benefits and sustainability of digital twins are multifaceted and deeply impactful. By enhancing resource efficiency, promoting environmental stewardship, fostering innovation, and supporting workforce development, digital twins not only provide immediate value but also contribute to a sustainable and prosperous future. As we continue to explore and refine these technologies, their potential to drive meaningful change in a wide array of industries will only grow, making digital twins an indispensable tool in our quest for a better tomorrow.

Chapter 25: Overcoming Barriers to Adoption

Bringing digital twins into the mainstream requires addressing a variety of challenges and hurdles that go beyond technology. The allure of digital twins lies in their potential to transform industries through streamlined processes, optimized operations, and dynamic simulations. Yet, barriers such as organizational inertia, high up-front investment costs, and complexities in integration present significant obstacles. These can be surmounted through a thoughtful strategy that aligns technology with business goals, making the intangible benefits of digital twins more concrete. Leaders need to inspire innovation by crafting compelling narratives that showcase the potential for return on investment and long-term advantages. Advocating for a shift in mindset, where digital twin technology is seen as indispensable as traditional tools, is crucial. Encouraging a culture of experimentation and learning within organizations can further reduce resistance. As we dismantle these barriers, we open pathways not just to adoption, but to robust innovation, setting the stage for digital twins to become pivotal enablers of the future digital landscape.

Addressing Hurdles and Challenges

Digital twins, while brimming with potential, often encounter significant hurdles on the path to widespread adoption. These challenges are not just technical but also organizational and cultural. To truly unlock the disruptive power of digital twins across industries, one must navigate these complexities with a strategic mindset. It's essential to understand and address the intricacies that can hinder progress, ranging from technological constraints to organizational inertia.

Technological barriers are perhaps the most prominent. Integration is key, as digital twins rely heavily on synthesizing diverse data sources, utilizing advanced simulation tools, and coordinating with existing systems. The technological landscape is frequently fragmented, with legacy systems that weren't designed to support such innovative solutions. As a result, companies face the daunting task of overhauling existing infrastructures without disrupting ongoing operations. Overcoming these barriers requires careful planning and often, substantial investment in new technologies that can bridge gaps between old and new.

Data quality and availability also pose significant bottlenecks. Digital twins are only as good as the data they consume, and poor data can lead to inaccurate models and suboptimal decisions. Ensuring data accuracy, consistency, and timeliness is crucial. This can be particularly challenging in industries where data is siloed or where it comes from a variety of sources with differing structures and standards. Organizations must invest in robust data management and governance frameworks that ensure data is clean, accessible, and usable.

Furthermore, there's the issue of scalability. Digital twin technology needs to handle massive volumes of data in real-time, which can strain existing IT resources. Companies may need to adopt cloud-based solutions to scale effectively without compromising performance. Cloud environments can offer the flexibility and computational power necessary to support the continuous growth of digital twin applications, yet migrating to the cloud is an investment of its own, demanding foresight and careful allocation of resources.

Culturally, resistance to change is a significant challenge. Implementing digital twins often requires a shift in organizational mindset and processes, which might be met with skepticism or resistance by employees. There can be a fear of job displacement or a hesitancy to learn new tools, which can stall adoption. Leaders must foster an environment where innovation is embraced and continuous learning is encouraged. Providing training and clear communication about the benefits and changes digital twins bring can alleviate fears and drive adoption.

Security and privacy concerns are another critical area that organizations cannot afford to overlook. Digital twins deal with vast amounts of sensitive and proprietary data. Protecting this data from cyber threats while ensuring compliance with regulatory requirements is paramount. Companies need to implement stringent cybersecurity measures and regularly update them to safeguard this data. Building trust with stakeholders about how data is handled and protected can go a long way in easing concerns surrounding data security.

Beyond organizational and technological barriers, there are also broader market challenges. The lack of industry standards for digital twins can lead to incompatibilities and fragmented solutions that hinder interoperability. Consortia and industry groups are working to develop and promote standards, but this is an ongoing effort. Companies need to keep abreast of these developments and work towards adopting best practices that align with emerging standards, which will foster interoperability and reduce fragmentation.

Finally, there is the challenge of proving the value of digital twins through measurable benefits. Many stakeholders are cautious about adopting cutting-edge technology without clear evidence of return on investment (ROI). Organizations must focus on demonstrating tangible outcomes from digital twin projects, such as cost savings, improved efficiency, or innovation opportunities that directly impact their bottom line. This can help in securing executive buy-in and wider organizational support, paving the way for successful implementation.

Overcoming these hurdles requires an integrated approach that blends technology with strategy, collaboration, and education. By addressing these challenges head-on, organizations can position themselves to fully capitalize on the potential of digital twins and drive meaningful change across industries. With concerted efforts, the hurdles to adoption can be transformed into stepping stones toward a future where digital twins are an integral part of the technological landscape.

Inspiring Innovation and Change

Inspire. Innovate. Transform. These are the pillars that drive the adoption of digital twins, despite numerous barriers. We live in an era where digital transformation is not optional but essential. It paves the way for groundbreaking improvements, reshaping industries while fostering new growth avenues. The potential of digital twins lies not just in enhancing operational efficiency but in enabling transformative change across both traditional and emerging sectors.

There's no denying that the road to adopting digital twins is fraught with challenges. However, innovation leaders must embrace this technology's potential to drive positive change. How then, can these trailblazers champion the cause? It begins with cultivating a culture of curiosity and openness. By fostering an environment where questioning the status quo is encouraged, organizations can unlock untapped potential and pioneer solutions that were once deemed impossible.

The journey of innovation often requires organizations to rethink their approaches fundamentally. It demands a keen understanding of the technology's intricacies balanced with a vision of its strategic applications. Sector by sector, digital twins offer unique advantages that could reshape the ground rules of competition and service delivery. For example, in manufacturing, adopting digital twins can lead to unprecedented levels of production efficiency and accuracy. In healthcare, they hold the promise of truly personalized medicine. Such examples serve as blueprints for other industries looking to innovate.

Curiosity and creative thinking are at the heart of innovation. Encouraging cross-disciplinary teams to collaborate and share insights can ignite inspiration needed to breach the walls of limitation. When industry professionals from diverse backgrounds come together, they challenge each other's perspectives and develop comprehensive, groundbreaking solutions. Organizations must leverage this synergy to refine their strategies and deploy digital twin technologies effectively.

Yet, fostering a culture of innovation isn't merely about encouraging ideas—it's about implementing them. Organizations must be ready to take calculated risks, as they can become the driving force for change. Success stories in various sectors underscore the significance of a bold approach. History reminds us that real innovation doesn't occur in a vacuum but thrives in an environment where traditional barriers are demolished for new pathways to be discovered.

Moreover, the role of digital twins in catalyzing change is analogous to the role of artists in crafting a masterpiece. Each line, each brushstroke, although seemingly minute, contributes to an awe-inspiring work of art. Similarly, digital twins allow organizations to painstakingly refine each process, each operation, and improve them continuously. This meticulous nature of digital twins provides a holistic perspective, enabling industries to innovate from the ground up.

Consider the energy sector where traditional power plants face efficiency and environmental footprint challenges. By adopting digital twins, they can simulate energy flows, optimize production, and minimize waste. Not just reactive, this technology's proactive nature instigates far-reaching systemic change, heralding a new age of sustainable energy solutions that balance resource management with environmental conservation.

Change rarely occurs without leadership. Visionary leaders and decision-makers play a crucial role in cultivating the conditions for innovation to flourish. These leaders must not only drive initiatives within their organizations but also advocate for broader systemic change. Be it through establishing partnerships with technology providers or investing in employee skill development, they ensure their organizations remain at the forefront of their respective fields.

Inspiring change requires a comprehensive understanding of the challenges faced by distinct organizations and industries. Tailored solutions crafted with a strong understanding of pain points ensure the smooth rollout of digital twin technologies. Organizations must foster a collaborative dialogue with stakeholders, gaining insights that can guide the development of disruptive solutions capable of transforming operational landscapes.

Further driving innovation is the interconnectedness offered by digital twins—syncing real and digital worlds into a seamless continuity. This capability allows industries to simulate, predict, and implement changes with confidence and precision. It is this feedback loop that enables businesses to evolve dynamically, adapting to market needs and capitalizing on real-time data insights.

Additionally, change is also about inspiring the workforce—a component often overlooked but essential for a digital twin's successful adoption. Empowering employees through training and development, fostering an environment that rewards innovative thinking, and valuing different perspectives ensures organizations have a team aligned with technological goals. When everyone is on board, the transformation gains momentum.

Adoption hurdles are real, but digital twins ask organizations to imagine what's possible, and as that vision expands, it kindles a drive to create the future. It is this forward-looking mindset that fuels innovation. An organization's willingness to leap forward despite uncertainties defines whether it leads or follows in this digital era. Successful cases have demonstrated that those who choose to lead inspire change, reshape industries, and leave a lasting legacy.

Ultimately, the transformative power of digital twins rests not solely on technology itself, but on a shared belief in what industry professionals can achieve. Here, inspiration begets integration, and innovation becomes a palpable reality, driving meaningful change. Behind every step forward lies a story of curiosity, challenge, and courage that pushes the limits of what's possible, ushering in an age of limitless potential through digital twins.

Conclusion

The transformative journey of digital twins reflects a broader movement towards a more integrated and intelligent world. As we've ventured into the realms of how digital twins influence various industries, it's clear that they stand on the verge of revolutionizing everything from healthcare and manufacturing to urban planning and energy management. Each sector unlocks new possibilities, enabling us to harness real-time data, simulate scenarios, and devise impeccable strategies for problem-solving and optimization.

We find ourselves at a crossroads, where technology is not merely a tool but a partner in innovation. Digital twins embody this synergy, offering a rich tapestry of opportunities for those willing to grasp them. They compel us to rethink traditional models and to leap into a future replete with dynamism and responsiveness. Engaging with these virtual counterparts, leaders and decision-makers can drive greater efficiency, unlock hidden insights, and innovate beyond current perceived limits. The digital twin is not just a reflection; it's an agent of change.

With the complexities of modern challenges, this duality between the digital and the physical becomes increasingly essential. Precision, foresight, and agility define successful organizations, and digital twins are instrumental in achieving these objectives. The potential for increased efficiency and resource optimization is immense. However, it is paramount to approach this evolution with a balanced mindset, considering ethical implications, data privacy, and security, which are pivotal as the technology scales.

Envisaging a future with pervasive digital twin adoption, the landscape of opportunities feels boundless. The pivotal role that digital twins will soon play in our daily lives requires us to build infrastructures and ecosystems that are robust, secure, and flexible. Embracing collaboration, both within industries and across them, can accelerate developments and broaden the impact of digital twins. This collaborative ethos will foster an environment conducive to sharing insights and driving cohesive innovation.

Much of the allure surrounding digital twins stems from their ability to not only predict but also to adapt and learn autonomously over time. With advancements in artificial intelligence and machine learning, digital twins will become increasingly sophisticated, evolving into proactive entities that do much more than mirror their physical counterparts. They'll offer predictive insights that can mitigate risks, prevent failures, and essentially transform proactive management into an industry standard. The reciprocal relationship between AI and digital twins holds boundless promise, reinforcing each other's potential in astounding ways.

Yet, realizing the full value of digital twins requires strategic foresight and a readiness to embrace change. For organizations, this means nurturing a culture of continuous learning and innovation, fostering an environment where interdisciplinary cooperation is encouraged. The digital twin era demands a workforce that is adept at navigating complex data landscapes and that possesses the curiosity to explore uncharted territories in technology. The success of digital twins hinges on not only the technology itself but on the human willingness to adapt and explore.

The narrative surrounding digital twins is inspirational, urging us to dream bigger and to act with intentionality. As we nurture this technology, it's essential to envision its role as more than a mere technological tool — it's an enabler of the future we aspire to build. From transforming industries to enhancing everyday interactions, the digital twin phenomenon reaffirms our faith in technology as a force for good growth and positive change. This burgeoning horizon beckons with the promise of improving lives, elevating business capabilities, and steering communities towards a sustainable future.

While there are still challenges to overcome, ranging from technical to ethical, these should not deter us but rather intensify our resolve to address them. By fostering open discourse and shared explorations, the digital twin community can pave a pathway that's both progressive and conscientious. The narratives we've explored depict the nascent stage of a journey filled with potential discoveries and innovations still unfolding on the canvas of technology.

In conclusion, the momentum behind digital twins is undeniable and their journey is an inspiring testament to human ingenuity. As we conclude this exploration, let it serve as a call to action: to participate actively, to innovate bravely, and to drive towards a future where technology and humanity coexist harmoniously. The road ahead offers unparalleled opportunities, inviting dreamers and doers alike to redefine the limits of what's possible.

Technical Glossary and Resource List

This section serves as a comprehensive glossary to support your journey into the world of digital twins, offering clear definitions and valuable resources tailored for technology enthusiasts, industry professionals, and decision-makers. In a rapidly evolving landscape, understanding jargon and accessing the right information is crucial for demystifying and championing digital twin technology.

Glossary

Digital Twin: A virtual representation of a physical object, process, or system, used for simulation, analysis, and control.

IoT (Internet of Things): Network of physical objects embedded with sensors, software, and other technologies to connect and exchange data with other devices and systems over the Internet.

AI (Artificial Intelligence): Machines' ability to mimic human intelligence, learning from experience and performing tasks typically requiring human intellect.

Predictive Maintenance: Technique for forecasting equipment failures and scheduling maintenance preemptively to prevent downtime and reduce costs.

Smart Cities: Urban areas that use various sensors and technologies to collect data for efficiently managing assets, resources, and services.

Cloud Computing: Delivery of computing services over the Internet, enabling flexible resources and economies of scale.

Augmented Reality (AR): Interactive experience where digital information is overlaid on the real world, enhancing perceptions and interactions.

Virtual Reality (VR): Use of computer technology to create a simulated environment, immersing the user in a virtual world.

Machine Learning: A type of AI that enables systems to learn from data patterns without being explicitly programmed.

Blockchain: A digital ledger technology for securely recording transactions across several computers to ensure data integrity and trust.

Integration: The process of combining different computing systems and software applications physically or functionally to act as a coordinated whole.

API (Application Programming Interface): A set of protocols and tools that allows different software applications to communicate with each other.

Data Privacy: Concerns relating to protecting personal information handled by organizations to ensure it is not misused.

Scalability: The capability of a system to handle a growing amount of work, or its potential to be enlarged to accommodate that growth.

Sustainability: The practice of maintaining processes or systems in ways that do not negatively impact social, economic, or environmental resources.

Resource List

Books: "Digital Twin: The New Era of Intelligent Operations" "Industry X.0: Realizing Digital Value in Industrial Sectors"

Online Platforms: MIT Technology Review (www.technologyreview.com) Gartner (www.gartner.com)

Conferences: Digital Twin World Conference IoT Tech Expo

Communities: Kaggle (www.kaggle.com) IEEE Digital Twin Initiative

As you continue exploring the potential of digital twins, these terms and resources will equip you with pertinent insights and practical tools, encouraging you to drive innovation and meaningful change in your field.

www.ingramcontent.com/pod-product-compliance
Lightning Source LLC
Chambersburg PA
CBHW051238050326
40689CB00007B/980
* 9 7 8 1 4 5 6 6 5 7 9 2 5 *